D0856716

Application of Metal Cutting Theory

APPLICATION OF METAL CUTTING THEORY

Fryderyk E. Gorczyca, P.E., C.Mfg.E.
Professor of Mechanical Engineering
Southeastern Massachusetts University
North Dartmouth, Massachusetts

Industrial Press Inc.

Library of Congress Cataloging-in-Publication Data

Gorczyca, Fryderyk E.
Application of metal cutting theory.

Includes bibliographies and index.
1. Metal-cutting. 2. Manufacturing processes.
I.Title.
TJ1185.G625 1987 671.5'3 87-3882
ISBN 0-8311-1176-3

INDUSTRIAL PRESS
200 Madison Avenue
New York, New York 10016

APPLICATION OF METAL CUTTING THEORY

2 4 6 8 9 7 5 3

Dedication

To Stasia, my devoted wife and efficient word processor, and to Diane and Ann Marie, my lovely daughters, whose encouragement provided the inspiration for this work.

If you can measure that of which you speak, and can express it by a number, you know something of your subject; but if you cannot measure it, your knowledge is meager and unsatisfactory.

Lord Kelvin

Preface

Application of Metal Cutting Theory is an introduction to the basic scientific and economic concepts that serve as the foundation of metal cutting theory. This textbook is intended for all who have contact with or an interest in metal cutting.

The approach used in this volume is one of illuminating theory by practical examples: theoretical aspects are applied in sample examples that emphasize numerical solutions. Some of these examples deal with economics and are formulated to lead the reader in the direction of the efficient use of machining factors.

Mathematical derivations are presented in detail and are reinforced by numerical illustrations. It is the intent of the text to present the theory in clear, simple terms that can be understood readily and can be applied easily.

Many texts on metal cutting fall into one of two categories: they are either very descriptive with little emphasis on numerical examples or they are very theoretical with little emphasis on practical applications. It is the purpose of this text to bridge the gap between these two approaches.

This volume is written to appeal to all educational levels where metal cutting or applied science is involved, and is a portion of the subject matter presented in a senior technical elective entitled, "Tool Engineering," offered to Engineering Technology students at Southeastern Massachusetts University. This book can also be used in Manufacturing Processes courses where an extended treatment of metal cutting is desired.

Since the text is written at an introductory level, it can be adopted to satisfy an Applied Science requirement for either Business or Liberal Arts programs. In a similar fashion, it can be adopted in a community college or technical school setting. In an elementary course presentation at a technical school level, the mathematical rigor associated with the analytical derivations can be omitted; instead, a concentration on the highlights of the relationships described in the derived equations can be stressed.

The text should also appeal to industry, especially where an understanding of the relationship between the tool and the workpiece is desired in light of the influence that machining factors have on the costs of a particular operation.

There are five chapters in this textbook. The first chapter deals with general economic considerations. Matters of concern that dictate conditions for the practical application of tooling are discussed. Topics include the effect that cutting speed has on production rate and on the cost of production; the production gains that can be attained through the use of a more efficient tool; the determination of cost break-even points; the effect of the size of production on the cost of tooling; and a discussion of tooling investment strategy.

The second chapter deals with the fundamentals of the metal cutting process. Basic scientific terms are introduced in this chapter along with experimental techniques for the evaluation of the physical properties of metals. Of special interest is the mechanical equivalent of heat, which is introduced with the appraisal of the temperature rise in the scrap chip of a metal cutting operation. This concept is viewed from the perspective of a simplified model.

The third chapter introduces the major cutting tool materials. Particular tooling applications are highlighted. Additional topics covered are the heat-treatment process; special tool surface treatments; specific tool coatings; and influence of the cutting fluid on the metal cutting operation.

The fourth chapter deals with the mechanics of the cutting process. The forces acting between the tool and the workpiece are scrutinized. These forces are further resolved into convenient components that are used to extract the characteristics of the cutting operation. Other topics included are the coefficient of friction; power consumption; motor horsepower; and effective clearance angles on tools.

The last chapter covers the topics of tool wear and affiliated production costs. Analytical expressions for tool wear and for conditions related to maximum production and to minimum cost are derived in this chapter. These optimizing techniques are then used to compare the performance of different tooling. It is pointed out that it is not as important to have good tooling available as it is to use the good tooling to advantage. A final topic in this chapter indicates the influence of financial considerations on effective tooling expenditures.

Since so many have contributed directly or indirectly to the contents of this text, the author expresses his gratitude to one and all for their assistance. To my students who have so eagerly compiled the experimental data used in the technical examples, I affirm my gratitude. To my colleagues who provided technical assistance, I voice my appreciation. To my friends in industry who provided information to make the illustrated examples realistic and current, I am deeply indebted. To the inspirational educators who were able so masterfully to ignite my curiosity in the subject matter, I offer a respectful thank you. Finally, to Stasia Gorczyca, my collaborator on this project, I extend my compliments for being able to transform loosely scribed notes to a neat and clean manuscript.

Contents

Chapter 3 Cutting Tool Materials

Chapter 4 Mechanics of the Cutting Process

Chapter 5 Tool Wear and Affiliated Production Costs

1
Economic Considerations

1.1 INTRODUCTION

Economic considerations are of prime importance when metal cutting theory is applied to a manufacturing process. The accurate appraisal of economic factors can ultimately dictate conditions for the most efficient use of a cutting tool. Proper setting of cutting speeds, feeds, and depth of cut, as well as the proper choice of a cutting tool material or of a cutting fluid, can lead to major economic improvements in a machining operation. Changes that may seem on the surface to be insignificant are capable, in many cases, of having a profound influence on costs of operation, especially in an application where high volume is involved.

In this chapter, a series of manufacturing topics are discussed in a broad fashion. The importance of economic analysis is emphasized with these evaluations. The first topic deals with the general relationship between the cutting speed setting and the rate of production. As the cutting speed is increased from low levels, a corresponding increase in production rates takes place. However, there is a point, that is, a cutting speed setting, that will result in a maximum production rate. Beyond this point, as the cutting speed increases, tool wear as well as machine downtime lead to diminishing returns. As a result, the production rate begins to dramatically decrease.

A similar illustration is given for the relationship between cutting speed and cost of production. At low speeds, production costs are obviously high since each part bears the burden of having a high machining time attached to it. As the cutting speed increases from low levels, costs tend to decrease since production rates per unit time are higher. This trend leads to a point of minimum cost operation. Beyond this point, as the cutting speed increases, costs begin to increase due to additional cost charges affiliated with tool wear and corresponding machine downtime.

Of special interest is the condition where the cutting speed is not set at either maximum production or at minimum cost. An analysis reveals that there are two different cutting speed settings that will yield the same production rate. Also, there are two different cutting speed settings that will yield the same cost of production. This situation justifies an appraisal of what may be considered the most favorable machine setting. This leads to a production decision that is influenced by either volume demands or by the cost of production. Illustrations of these cases are given with numerical examples.

The use of a more efficient tool is also discussed in this chapter. Ordinarily, a more efficient tool possesses characteristics that enable it to produce more parts per unit time, with a corresponding reduction in machine downtime. A tool of this type increases production through higher operating speeds, which lead to an increase in the efficiency of the machine on which the tool is being used.

Many times the decision to choose a more efficient tool is not restricted to a single tool choice. If production and tool performance data are known for a number of different tools, then charts of the break-even variety can prove to be valuable aids when making the proper tooling selection for an operation. These important decisions can be justified through analysis. Two techniques are available. One deals with an analytical approach using mathematically derived equations, whereas the other approach deals with a simplified graphical solution.

Additional topics covered in this chapter include the limitations on tooling investments for a fixed production run; the relationship of the tooling cost to the size of production; and a discussion of an investment strategy for potential repeat orders. In all cases, analytical expressions are derived while explaining the important relationships. Finally, numerical examples are given to reduce the theory to practice. The objective of this chapter is to point out in a broad general sense a series of cases where a shrewd accurate analysis can lead to an understanding of the factors that control an efficient metal cutting operation.

1.2 PRODUCTION RATE AND CUTTING SPEED

The relationship between production rate and cutting speed can be generalized by a graph, as illustrated in Fig. 1.1. At point 1, there is no production because the cutting speed is zero. As the cutting speed is increased, there is an obvious increase in production, as

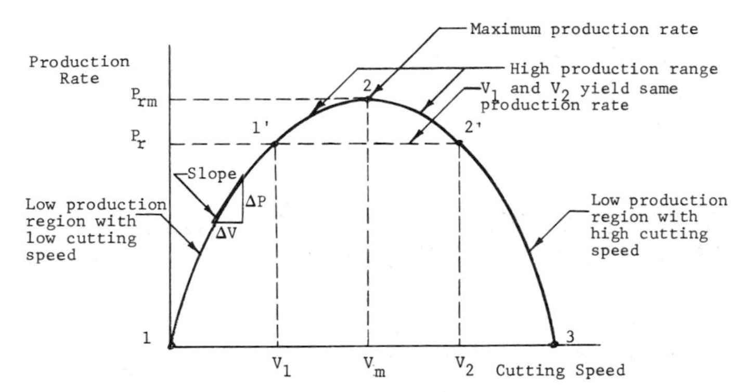

Fig. 1.1. Production rate versus cutting speed.

illustrated by the curve span 1–1′, a region of relative low production that is directly affiliated with a low cutting speed range. The curve span 1′–2–2′ is a region of relative high production. Of special interest is point 2, where the slope of the curve, $\Delta P/\Delta V$, is zero, that is, a line drawn tangent to the curve at this point will be horizontal. This point identifies a cutting speed, V_m, that can be labeled as the idealized cutting speed that produces the maximum production rate P_{rm}.

As the cutting speed is increased beyond point 2, production begins to decrease. The curve span 2–3 represents diminishing returns. As can be seen in Fig. 1.1, increases in the cutting speed beyond point 2, that is, V greater than V_m, tend to reduce the production rate. This is usually due to tool wear, resulting in downtime in the operation, during which the machine is not operating and the corresponding time is used to replace and adjust the tool. Point 3 is of interest since it depicts a high cutting speed that causes a tool failure on contact. The production rate at point 3 is obviously zero. At this point, an instantaneous catastrophic failure occurs.

Points 1′ and 2′ characterize the same production rate, P_r, for two different cutting speeds, V_1 and V_2. If the machine setting is not at the idealized cutting speed, V_m, then there are always two cutting speeds that will deliver the same production rate. The reason this takes place is that usually there is a compromise between the lower speed, V_1, and the higher speed, V_2. At the lower speed, production is sensitized to machine time, whereas at the higher speed, production is sensitized to the machine downtime related to replacing worn-out tools. The machine is operating slower at point 1′, resulting in production rate P_r. The machine is operating faster at point 2′ and produces more while it is operating, but, owing to higher tool wear, machine downtime is increased. This condition demands attention and results in consuming nonproductive time to change the tools. The result is the same production rate P_r.

1.3 EXPERIMENTAL DETERMINATION OF PRODUCTION—CUTTING SPEED CURVE

A test can be conducted to determine the relationship between production rate and cutting speed. By measuring the production rate as the cutting speed is varied, a curve of the type shown in Fig. 1.1 can be compiled. Let us assume that a test of this type has been conducted. The production rate (parts produced per hour) has been recorded as the cutting speed (feet per minute) has been varied in increasing increments of 50 ft/min. The results of the test are listed in Table 1.1.

A plot of the experimental data is given in Fig. 1.2.

If an assumption is made that the curve represented in Fig. 1.2 is parabolic in shape, then an equation can be derived to analytically express the relationship between production rate and cutting speed. An equation of the form

$$P = AV^2 + BV + C \tag{1.1}$$

(where P is the production rate, V is the cutting speed, and A, B, and C are constants) can conveniently be fitted to pass a parabolic curve through any three given points. For convenience, let us use the three points of (V, P) as (0, 0), (200, 100), and (400, 0). These are points 1, 2,

Table 1.1. Experimental Data of Production Rate and Cutting Speed

	Cutting Speed		Production Rate
Test	(ft/min)	(m/min)	(parts/hr)
1	0	0	0
2	50	15.24	43.75
3	100	30.48	75
4	150	45.72	93.75
5	200	60.96	100
6	250	76.20	93.75
7	300	91.44	75
8	350	106.68	43.75
9	400	121.92	0

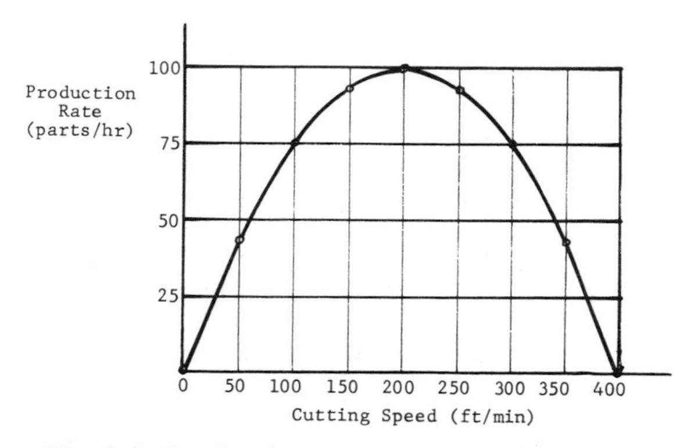

Fig. 1.2. Production rate versus cutting speed.

and 3 as listed in Fig. 1.1. The constants of Eq. 1.1 can be evaluated by substituting values of V and P.

For $V = 0$ and $P = 0$, Eq. 1.1 can be written as

$$0 = A(0)^2 + B(0) + C$$

from which $C = 0$. For $V = 200$ and $P = 100$, Eq. 1.1 can be written as

$$100 = A(4 \times 10^4) + B(200) \qquad \textbf{(1)}$$

For $V = 400$ and $P = 0$, Eq. 1 can be written as

$$0 = A(16 \times 10^4) + B(400) \qquad \textbf{(2)}$$

Multiplying Eq. 2 by $-\frac{1}{2}$ and adding Eqs. 1 and 2 enables the isolation of A. As a result,

$$A = -2.5 \times 10^{-3}$$

Substituting A into either Eq. 1 or Eq. 2 enables the constant B to be determined. As a result,

$$B = 1$$

Equation 1.1 now can be written as

$$P = -2.5 \times 10^{-3} V^2 + V \qquad \textbf{(1.2)}$$

The data in Table 1.1 were calculated from Eq. 1.2.

EXAMPLE 1.1. Determine the two cutting speeds that will result in a production rate of 85 parts/hr if a manufacturing operation is described by Eq. 1.2.

Substituting into Eq. 1.2 yields

$$85 = -2.5 \times 10^{-3} V^2 + V$$

or

$$2.5 \times 10^{-3} V^2 - V + 85 = 0 \tag{1}$$

Equation 1 is a general quadratic equation of the form

$$AX^2 + BX + C = 0$$

where $A = 2.5 \times 10^{-3}$, $B = -1$, $C = 85$, and $X = V$. The solution of this equation is given by the quadratic formula in the form

$$X = \frac{-B \pm \sqrt{B^2 - 4AC}}{2A} \tag{1.3}$$

Solving Eq. 1 by substituting in Eq. 1.3 yields

$$V = 122.54,\ 277.46 \text{ ft/min } (37.35,\ 84.57 \text{ m/min})$$

EXAMPLE 1.2. The results of three tests conducted on the relationship between production rate and cutting speed are:

Test 1: $V_1 = 100, \quad P_1 = 8$

Test 2: $V_2 = 200, \quad P_2 = 12$

Test 3: $V_3 = 300, \quad P_3 = 10$

(a) Assuming a parabolic relationship between the test points, derive an expression describing the production rate as a function of cutting speed.

(b) Solve for the cutting speed that will give a maximum production rate.

Substituting into Eq. 1.1 yields

$$8 = A(10^4) + B(100) + C \tag{1}$$

$$12 = A(4 \times 10^4) + B(200) + C \tag{2}$$

$$10 = A(9 \times 10^4) + B(300) + C \tag{3}$$

Solving Eq. 1 for C and substituting into Eqs. 2 and 3 yields

$$4 = A(3 \times 10^4) + B(100) \tag{4}$$

$$2 = A(8 \times 10^4) + B(200) \tag{5}$$

Multiplying Eq. 4 by -2 and adding Eqs. 4 and 5 yield

$$A = -3 \times 10^{-4}$$

Substituting A into Eq. 4 yields

$$B = 0.13$$

Substituting A and B into Eq. 1 finally gives

$$C = -2$$

Equation 1.1 can now be written as

$$P = -3 \times 10^{-4}V^2 + 0.13V - 2 \tag{6}$$

To find the maximum production cutting speed, Eq. 6 is differentiated and the slope is set at zero. As a result,

$$\frac{dP}{dV} = -6 \times 10^{-4}V + 0.13 = 0$$

Solving for V yields

$$V = 216.7 \text{ ft/min } (66.05 \text{ m/min})$$

An explanation of the differentiation process is given as follows. The derivative or slope of a line tangent to the curve is defined as the limit of the slope $(\Delta P/\Delta V)$ when ΔV approaches zero or

$$\lim_{\Delta V \to 0} \frac{\Delta P}{\Delta V} = \frac{dP}{dV}$$

It can be derived as follows. From Eq. 6 at a point (P, V)

$$P = -3 \times 10^{-4}V^2 + 0.13V - 2 \tag{7}$$

At some other point $(P + \Delta P, V + \Delta V)$ where ΔP and ΔV are small increments, Eq. 6 can be written as

$$P + \Delta P = -3 \times 10^{-4}(V + \Delta V)^2 + 0.13(V + \Delta V) - 2 \tag{8}$$

Subtracting Eq. 7 from Eq. 8 gives

$$\Delta P = 6 \times 10^{-4}V(\Delta V) - 3 \times 10^{-4}\Delta V^2 + 0.13\Delta V$$

from which

$$\frac{\Delta P}{\Delta V} = -6 \times 10^{-4}V - 3 \times 10^{-4}\Delta V + 0.13$$

Since

$$\frac{dP}{dV} = \lim_{\Delta V \to 0} \frac{\Delta P}{\Delta V} = \lim_{\Delta V \to 0} (-6 \times 10^{-4}V - 3 \times 10^{-4}\cancelto{0}{\Delta V} + 0.13)$$

then

$$\frac{dP}{dV} = -6 \times 10^{-4}V + 0.13 = \text{slope of the curve}$$

1.4 COST OF PRODUCTION AND CUTTING SPEED

The relationship between cost of production and cutting speed can be generalized by a graph, as illustrated in Fig. 1.3. Point 1 represents a high cost of production resulting from a slow cutting speed. This condition requires a relatively large amount of time to produce the part. As the cutting speed is increased from point 1, costs are reduced as less time is required to produce each part. The curve span 1–2 dramatizes the effects of cost reduction through increased productivity. If all manufacturing influencing factors remained constant, the span 1–2 would be a linear relationship represented by a straight line. However, the slope of the curve begins to decrease as the cutting speed is increased, indicating that factors related to tool wear are having a profound influence. At point 2, a cutting speed V_n yields the ideal minimum cost of production.

Beyond this point, diminishing returns produce higher costs of operation. The curve span 2–3 indicates rising costs that are affiliated with increases in the cutting speed. This is due to inefficiencies that influence the operation as the speed is increased. These may be summarized in terms of increased cost of tooling due to wear as well as the increased cost that must be absorbed due to downtime that is required to change the worn out tools.

Of special interest are points 1′ and 2′. These points represent two different speeds that result in the same cost of production labeled as C_v. Although points 1′ and 2′ indicate cutting speeds that result in the same unit cost per part, a further analysis may reveal that one of these points has an advantage over the other, that is, production evaluation may indicate that with point 2′ operations, more parts per unit time can be produced. In comparing a balance between point 1′ and point 2′ operations, on one side is low speed, V_1, reflecting low tool cost

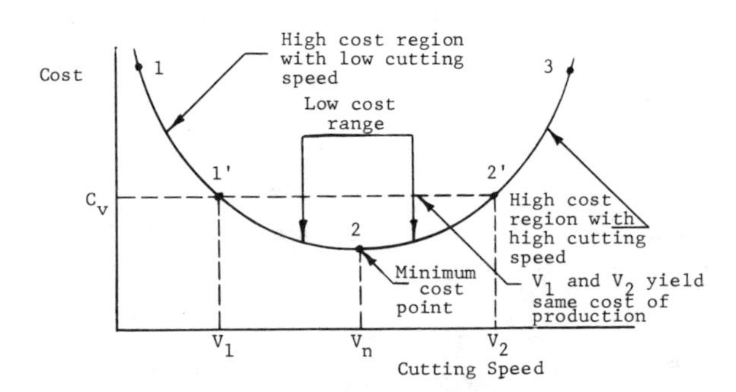

Fig. 1.3. Cost versus cutting speed.

but high machining cost. On the other side is high speed, V_2, reflecting high tool cost and high downtime cost due to high tool wear and the corresponding time required to change the tool.

1.5 EXPERIMENTAL DETERMINATION OF COST OF PRODUCTION—CUTTING SPEED CURVE

A test can be conducted to determine the relationship between cost of production and the cutting speed. By evaluating the cost of production as a function of the cutting speed, a curve of the type shown in Fig. 1.3 can be compiled. Let us assume that a test of this type has been conducted. The cost of production (in dollars) per unit produced has been documented as the cutting speed (feet per minute) has been varied in increasing increments of 50 ft/min. The results of the test are listed in Table 1.2.

A plot of the experimental data is graphically displayed in Fig. 1.4. A parabolic expression of the form

$$C_p = AV^2 + BV + C \tag{1.4}$$

(where C_p is the cost of production, V is the cutting speed, and A, B, and C are constants) can be fitted to the curve in Fig. 1.4 by choosing three points of convenience. Let us take the three points of (V, C_p) as (50, 10.02), (200, 3.17), and (300, 8.98). Substituting into Eq. 1.4 yields

$$10.02 = A(2500) + B(50) + C \tag{1}$$

$$3.17 = A(40{,}000) + B(200) + C \tag{2}$$

$$8.98 = A(90{,}000) + B(300) + C \tag{3}$$

Table 1.2. Experimental Data of Cost of Production and Cutting Speed

Test Number	Cutting Speed (ft/min)	Cutting Speed (m/min)	Cost of Production ($)
1	50	15.24	10.02
2	100	30.48	5.66
3	150	45.72	3.37
4	200	60.96	3.17
5	250	76.20	5.04
6	300	91.44	8.98
7	350	106.68	15.00

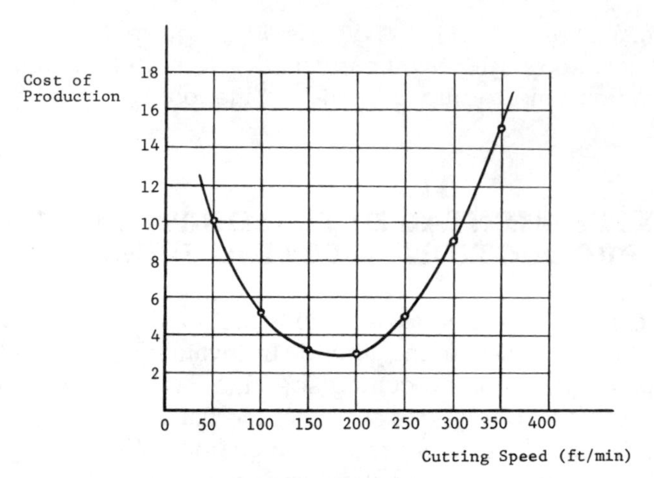

Fig. 1.4. Cost versus cutting speed.

Solving Eq. 1 for C and substituting in Eqs. 2 and 3 yields

$$-6.85 = A(37{,}500) + B(150) \quad \textbf{(4)}$$

$$-1.04 = A(87{,}500) + B(250) \quad \textbf{(5)}$$

Multiplying Eq. 4 by $-250/150$ and adding Eqs. 4 and 5 isolate A. As a result,

$$A = 4.151 \times 10^{-4}$$

Substituting the numerical value of A into Eq. 4 enables B to be evaluated. As a result,

$$B = -1.492 \times 10^{-1}$$

Finally, by substituting A and B into Eq. 1, C can be isolated:

$$C = 16.443$$

Equation 1.4 can now be written as

$$C_p = 4.151 \times 10^{-4} V^2 - 0.1492V + 16.443 \quad \textbf{(1.5)}$$

EXAMPLE 1.3. (a) Determine the cutting speed that will give the lowest cost of production, assuming the manufacturing process conforms with Eq. 1.5.

(b) Solve for the lowest cost of production from the data in part (a).

Differentiating (finding the slope of a line tangent to the curve) Eq. 1.5 yields

$$\frac{dC_p}{dV} = 8.302 \times 10^{-4} V - 0.1492$$

This operation can conveniently be performed from the formula

$$Y = CX^n$$

$$\frac{dY}{dX} = nCX^{n-1}$$

where C is a constant and n is an integer. Note that the terms in Eq. 1.5 are of the form X and Y listed above. Minimum cost exists when the slope of the curve is equal to zero or

$$8.302 \times 10^{-4}V - 0.1492 = 0$$

from which

$$V = 179.7 \text{ ft/min (54.8 m/min)}$$

Solving for the corresponding cost by substituting in Eq. 1.5 yields

$$C_p = 3.037 \text{ dollars}$$

EXAMPLE 1.4. Calculate the two cutting speeds that will provide a cost of production of $4.00/part for the process described by Eq. 1.5.

Substituting into Eq. 1.5 yields

$$4 = 4.151 \times 10^{-4}V^2 - 0.1492V + 16.443$$

or

$$4.151 \times 10^{-4}V^2 - 0.1492V + 12.443 = 0$$

Solving for V by using the quadratic formula Eq. 1.3 yields

$$V = 179.7 \pm 48.18$$

$$V = 131.52, 227.9 \text{ ft/min (40.09, 69.46 m/min)}$$

EXAMPLE 1.5. In an attempt to appraise the cost-cutting speed relationship, the following three tests were conducted:

Test 1: $V_1 = 85, \quad C_p = 2.25$

Test 2: $V_2 = 125, \quad C_p = 1.80$

Test 3: $V_3 = 170, \quad C_p = 2$

Assuming a parabolic relationship, write an equation that describes the operation between the given points.

From Eq. 1.4,

$$2.25 = A(85)^2 + B(85) + C \quad (1)$$

$$1.80 = A(125)^2 + B(125) + C \quad (2)$$

$$2.00 = A(170)^2 + B(170) + C \quad (3)$$

Solving Eq. 1 for C and substituting into Eqs. 2 and 3 yield

$$-0.45 = A(8400) + B(40) \tag{4}$$

$$-0.25 = A(21{,}675) + B(85) \tag{5}$$

Multiplying Eq. 4 by $-85/40$ and adding Eqs. 4 and 5 isolate A. As a result,

$$A = 7.438 \times 10^{-4}$$

Evaluating B by substituting the numerical value for A into Eq. 4 leads to

$$B = -1.675 \times 10^{-1}$$

The constant C is finally evaluated by substituting the numerical values for A and B into Eq. 1:

$$C = 11.114$$

As a result,

$$C_p = (7.438 \times 10^{-4})V^2 - (1.675 \times 10^{-1})V + 11.114$$

1.6 EVALUATION OF TWO CUTTING SPEEDS YIELDING THE SAME PRODUCTION RATE

In examining a metal cutting operation on a production machine, one can generalize conditions in terms of the machine production rate and the associated time affiliated with the operation. What must be taken into account is the time used while the machine is cutting as well as the time used when the machine is not cutting. A convenient expression for the production rate can be written as

$$P_r = \frac{M_r(T_t - T_d)}{T_t} \tag{1.6}$$

where

$$T_o = T_t - T_d$$

The terms are described as follows:

P_r is the production rate (parts/hr)
M_r is the machine rate while running (parts/hr)
T_t is the total time utilized in operation
T_d is the downtime encountered in operation
T_o is the operating time utilized in operation

In a similar fashion, the cost can be expressed conveniently as

$$C = M_c + D_c + T_c \qquad (1.7)$$

where C is the total cost per part, M_c is the machining cost per part, D_c is the downtime cost per part, and T_c is the tool cost per part. The following illustration is given to demonstrate the changes that take place when two different cutting speeds yield the same production rate. Let us assume that the machine speed was set at a value of V for case 1 and that it was increased to $1.5V$ for case 2. For a measured hour, the downtime was 5 min for case 1 and it increased to 23.33 min for case 2. The data are listed in Table 1.3.

Calculations for the entries in Table 1.3 are as follows: Solving for the machine rate for V gives

$$M_{r1} = \text{Parts produced/operating time}$$

$$M_{r1} = P_{r1}/T_{o1} = 10/55$$

$$M_{r1} = 0.1818 \text{ part/min} = 10.908 \text{ parts/hr}$$

In a similar fashion, the machine rate for $1.5V$ can be written as

$$M_{r2} = P_{r2}/T_{o2} = 10/36.67$$

$$M_{r2} = 0.2727 \text{ parts/min} = 16.362 \text{ parts/hr}$$

In examining the data of case 2 in Table 1.3, one may be inclined to draw the conclusion that if the downtime is reduced below 23.33 min/hr, then the recommendation to operate at $1.5V$ would be a sound one. In making an appraisal of this type, one needs to be cautious insofar as operation at $1.5V$ may produce more parts per unit time, but the consequence may be that these parts are more expensive to produce.

Case 2 in Table 1.3 documents that it is more costly to operate at $1.5V$. The results were calculated from the conditions that the production rate was 10 parts/hr with a machine charge of \$30.00/hr or \$0.50/min

Table 1.3. Listing of Experimental Data

Case	Cutting Speed (ft/min)	Production Rate (parts/hr)	Total Time (min)	Operating Time (min)	Downtime (min)	Machine Rate (parts/hr)
1	V	10	60	55	5	10.908
2	$1.5V$	10	60	36.67	23.33	16.362

Case	Total Cost (\$/part)	Machining Cost (\$/part)	Downtime Cost (\$/part)	Tool Cost (\$/part)
1	3.14	2.75	0.25	0.14
2	3.27	1.83	1.17	0.27

and with a tool wear analysis revealing that the tool cost is 5% of machining cost at the lower speed, V, and 15% of machining cost at the higher speed, $1.5V$.

Under these circumstances, it can be written for case 1 that

$$C_1 = M_{c1} + D_{c1} + T_{c1}$$

where

$$M_{c1} = \text{Machine time per hour per part} \times \text{charge rate}$$

$$M_{c1} = 55 \text{ min/hr} \times \text{hr}/10 \text{ parts} \times \$0.50/\text{min}$$

$$M_{c1} = \$2.75/\text{part}$$

$$D_{c1} = \text{Downtime per hour per part} \times \text{charge rate}$$

$$D_{c1} = 0.5 \times 0.5$$

$$D_{c1} = \$0.25/\text{part}$$

$$T_{c1} = 0.05 \times M_{c1}$$

$$T_{c1} = \$0.14/\text{part}$$

As a result,

$$C_1 = 2.75 + 0.25 + 0.14$$

$$C_1 = \$3.14/\text{part}$$

In a similar fashion, it can be written for case 2 that

$$C_2 = M_{c2} + D_{c2} + T_{c2}$$

where

$$M_{c2} = 3.667 \times 0.5$$

$$M_{c2} = \$1.83/\text{part}$$

$$D_{c2} = 2.333 \times 0.5$$

$$D_{c2} = \$1.17/\text{part}$$

$$T_{c2} = 0.15 \times 1.83$$

$$T_{c2} = \$0.27/\text{part}$$

As a result,

$$C_2 = 1.83 + 1.17 + 0.27$$

$$C_2 = \$3.27/\text{part}$$

Comparing cases 1 and 2 for cost per part reveals a \$0.13 differential. This indicates that it costs more to produce the parts at the

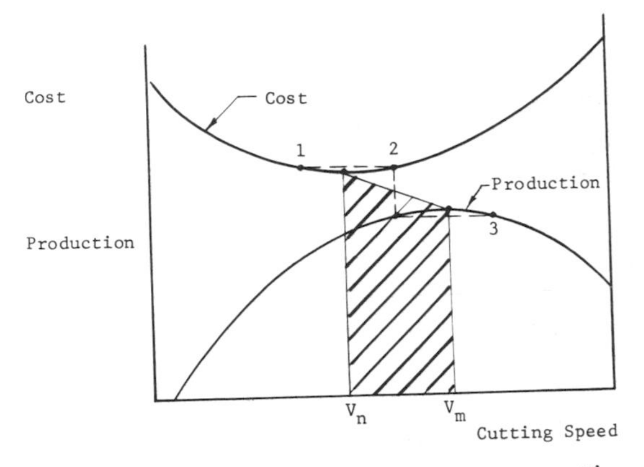

Fig. 1.5. Combined production and cost curves versus cutting speed range.

V_n and V_m do not coincide, a zone of operation between minimum cost and maximum production exists. The shaded area in Fig. 1.5 represents this zone. It is advantageous to set speeds in this range depending on a specific preference toward either lower costs or higher production. The dashed lines in Fig. 1.5 illustrate this choice.

If a cost criterion dictates a speed setting, then there is a choice between points 1 and 2. Point 2 obviously has an advantage since it enables more parts per unit time to be produced. On the other hand, if a production criterion dictates a speed setting, then point 2 again has an advantage over point 3 insofar as it provides for an operation with a corresponding lower cost with the same production as at point 3.

EXAMPLE 1.8. For the manufacturing process where two speed settings result in the same cost of production, the following data were compiled.

Eq. 1.7:	$C = M_c + D_c + T_c$
Case 1:	$3.00 = 2.50 + 0.35 + 0.15$
Case 2:	$3.00 = 2.27 + 0.49 + 0.24$

Assuming that the relationships in Eqs. 1.8–1.10 describe this operation, determine:

(a) The second speed setting if $V_1 = 200$ ft/min (60.96 m/min).
(b) The tool life for the second speed setting if the tool life for 200 ft/min was 45 min.

higher machine setting. A further examination shows that the $30.00/hr machine charge is distributed between the machining cost and the downtime cost. This takes place for the special case where the production rate is the same for the two speed settings. The difference in the cost of production in this example is due to the additional tooling cost generated at the higher speed.

EXAMPLE 1.6. Determine the machine rate for the case where a machine operates 90% of the time and produces 35 parts/hr.

From Eq. 1.6,

$$M_r = \frac{P_r}{T_o/T_t}$$

where $T_o/T_t = 0.9$. Substituting numerical values yields

$$M_r = \frac{35}{0.9} = 38.9 \text{ parts/hr}$$

EXAMPLE 1.7. A manufacturing process is analyzed to possess the following factors:

1. Production rate 35 parts/hr.
2. Machine rate $50.00/hr.
3. Tool depreciation cost 5% of machining cost.
4. Machine operating rate 90%.

From the given data, calculate the cost per part.

From Eq. 1.7,

$$C = M_c + D_c + T_c$$

where

$$M_c = (0.9/35) \times 50 = \$1.286$$

$$D_c = (0.1/35) \times 50 = \$0.143$$

$$T_c = 0.05 \times 1.286 = \$0.0643$$

Thus

$$C = \$1.49/\text{part}$$

1.7 EVALUATION OF TWO CUTTING SPEEDS YIELDING THE SAME COST OF PRODUCTION

Let us now consider an examination of a manufacturing process where a machine cutting speed setting of V_1 results in a cost of production of $2.00/part. This includes a machining cost of $1.65, a maintenance downtime cost of \$0.20, and a tool depreciation cost of \$0.15. An experimental trial notes that a 20% increase in the cutting speed reduces the tool life by 44%. It is desired to determine the cost to produce a part at the second speed setting where $V_2 = 1.2V_1$.

From Eq. 1.7, it can be written that

$$C_2 = M_{c2} + D_{c2} + T_{c2}$$

where

$$M_{c2} = \frac{M_{c1}}{V_2/V_1} \tag{1.8}$$

$$= \frac{1.65}{1.20} = \$1.38/\text{part}$$

and

$$D_{c2} = \frac{D_{c1}}{(1 - R_p)} \tag{1.9}$$

$$= \frac{0.20}{(1 - 0.44)} = \$0.357/\text{part}$$

Note: R_p = percent reduction in tool life. The assumption is made that the same percent reduction in tool life is reflected in a corresponding increase in downtime.

$$T_{c2} = \frac{T_{c1}}{(1 - R_p)} \tag{1.10}$$

$$= \frac{0.15}{0.56} = \$0.268/\text{part}$$

The cost for speed setting $1.2V_1$ is then

$$C_2 = \$1.375 + \$0.357 + \$0.268$$
$$= \$2.00/\text{part}$$

This is the same as the cost for speed setting V_1.

This example was concocted to have the same cost of production as for setting V_1. Despite the fact that V_1 and V_2 speed settings generate the same cost per part, a further examination of the data reveals that the V_2 setting will produce more parts per unit time. To illustrate, let R_m equal the machine charge rate for both V_1 and V_2. The time to machine one part for the low speed setting V_1 can be written as

$$t_1 = \frac{M_{c1} + D_{c1}}{R_m} \tag{1.11}$$

or

$$t_1 = \frac{1.65 + 0.20}{R_m} = \frac{1.85}{R_m}$$

On the other hand, the time to machine one part setting V_2 can be written as

$$t_2 = \frac{M_{c2} + D_{c2}}{R_m}$$

or

$$t_2 = \frac{1.37 + 0.357}{R_m} = \frac{1.627}{R_m}$$

Since the time required to machine one part is less for V_2, as seen from the above analysis, then there must obviously be an in[...] in the production rate. This increase can be measured by comp[...] the time required to machine one part. As a result, the increase in production rate from the low speed setting V_1 to the high speed setti[...] V_2 can be written as

$$\Delta P_r = \frac{P_2 - P_1}{P_1} = \frac{t_1 - t_2}{t_2} \tag{1.12}$$

$$\Delta P_r = \frac{1.85 - 1.627}{1.627} = 0.137 = 13.7\%$$

As can be seen, for the two machine speed settings that yield the same cost of production, the higher speed V_2 will produce 14% more parts. A conclusion can be reached that it is advantageous to operate at the higher speed for this particular cost analysis since the higher speed enables more parts to be produced per unit time. Table 1.4 lists the data for this example.

Figure 1.5 contains an example of combined production and cost curves for a wide cutting speed range. In examining the relationship of the cost and production curves, an assessment can be made that favors an efficient operation. Since the ideal cutting speed conditions

Table 1.4. Listing of Data

Case	Cutting Speed (ft/min)	Cost (\$/part)	Machining Cost (\$/part)	Downtime Cost (\$/part)	Tool Cost (\$/part)	Production Rate (parts/hr)
1	V	2.00	1.65	0.20	0.15	P_r
2	$1.2V$	2.00	1.375	0.357	0.268	$1.14P_r$

From Eq. 1.8,

$$V_2 = \frac{M_{c1} V_1}{M_{c2}} = \frac{2.5(200)}{2.27} = 220 \text{ ft/min } (67.06 \text{ m/min})$$

From Eq. 1.10,

$$R_p = 1 - \frac{T_{c1}}{T_{c2}} = 1 - \frac{0.15}{0.24} = 0.375, \text{ or } 37.5\%$$

Tool life can be written as

$$\mathrm{TL}_2 = \mathrm{TL}_1(1 - R_p) = 45 \times 0.625 = 28.13 \text{ min}$$

EXAMPLE 1.9. If case 1 in example 1.8 had a production rate of 50 parts/hr, what would be the production rate for case 2?

From Eq. 1.12,

$$\Delta P_r = \frac{t_1 - t_2}{t_2}$$

where

$$t_1 = \frac{2.50 + 0.35}{R_m} = \frac{2.85}{R_m}$$

$$t_2 = \frac{2.27 + 0.49}{R_m} = \frac{2.76}{R_m}$$

Thus,

$$\Delta P_r = \frac{2.85 - 2.76}{2.76} = 0.0326 \text{ or } 3.26\%$$

As a result,

$$P_r = R_{r1}(1 + \Delta P_r) = 50(1.0326) = 51.63 \text{ parts/hr}$$

1.8 USE OF A MORE EFFICIENT TOOL

A general definition of a more efficient tool is that it is not only one that produces more parts per unit time while operating, but it is also one that results in less downtime between tool changes. It is usually a tool that can operate at a higher speed and generates a lower downtime in the process. Figure 1.6 displays graphically the production rate of a so-called more efficient tool (tool 2) when compared to a reference tool (tool 1).

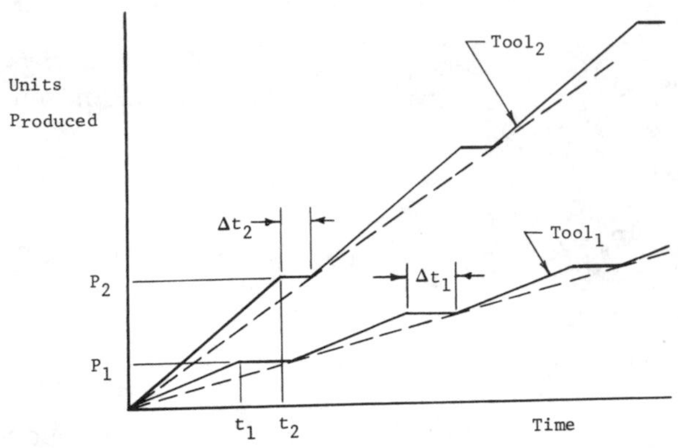

Fig. 1.6. Comparison of production rate and downtime of two different tools.

As can be seen in Fig. 1.6, tool 1 while operating, produces parts at a rate of P_1/t_1, whereas tool 2 produces parts at a rate of P_2/t_2. The corresponding downtime is designated as Δt_1 and Δt_2, respectively. Of interest is a means of measuring the gain that can be achieved by incorporating a more efficient tool in a given operation. The actual production rate represented by the slope of the dashed line in Fig. 1.6 can be written as

$$P_a = \frac{P_1}{t_1 + \Delta t_1} \tag{1.13}$$

where P_1 is the production during specific period of time, t_1 is the specific time period associated with machining operation, and Δt_1 is the downtime period. Of special interest when evaluating a more efficient tool is the anticipated increase in production. A percentage increase in production when comparing the performance of two tools can be expressed as

$$\Delta\% = \left(\frac{P_{a2} - P_{a1}}{P_{a1}}\right) \times 100 \tag{1.14}$$

where $\Delta\%$ is the percentage increase in production, P_{a1} is the actual production rate for tool 1, and P_{a2} is the actual production rate for tool 2.

The following illustration demonstrates how a more efficient tool can be evaluated for increases in production. Let us say that a given tool produces parts at a rate of 100 parts in a 2-hr period while operating and requires a machine downtime of 20% of the machining time. A more efficient tool is tested to operate satisfactorily at a rate of 150 parts in a 2.5-hr period while operating and requires a machine downtime of 10% of the machining time.

Solving for the actual production rate by substituting into Eq. 1.13 yields

$$P_{a1} = \frac{100}{2 + 0.4} = 41.67 \text{ parts/hr}$$

$$P_{a2} = \frac{150}{2.5 + 0.25} = 54.5 \text{ parts/hr}$$

From Eq. 1.14, the anticipated production rate percentage increase can be written as

$$\Delta\% = \left(\frac{54.5 - 41.67}{41.76}\right) \times 100 = 30.8\% \text{ increase}$$

The data of the previous example are listed in Table 1.5.

EXAMPLE 1.10. A production test of two tools reveals the following results from an 8-hr test period.

Tool	Number of Parts Produced	Period of Time (hr)	Corresponding Downtime (hr)
1	80	8	1
2	90	8	1.5

Calculate the production rate while the machine is in operation for each case as well as the production rate percentage increase of tool 2 over tool 1:

$$P_{r1} = \frac{P_1}{t_1} = \frac{80}{7} = 11.43 \text{ parts/hr}$$

$$P_{r2} = \frac{P_2}{t_2} = \frac{90}{6.5} = 13.85 \text{ parts/hr}$$

From Eq. 1.14,

$$\Delta\% = \left(\frac{90 - 80}{80}\right) \times 100 = 12.5\% \text{ increase}$$

Table 1.5. More Efficient Tool Data

Tool	Production Rate (parts/hr)	Actual Production Rate (parts/hr)	Downtime (hr/hr)	Production Rate Increase (%)
1	50	41.67	0.20	Reference
2	60	54.5	0.10	+30.8

1.9 BREAK-EVEN POINTS

In many cases, a more efficient tool is affiliated with a higher cost for the tool. This brings up the question, when is it profitable to use a more expensive tool? The question can be answered by an analysis of break-even points. To find the break-even point for a given tool application, it can be written that

Cost of operation with new tool = Cost of operation with old tool

This expression can be expanded to contain the particulars of the operation, that is,

$$N_n t_n R_n + C_t = N_o t_o R_o \tag{1.15}$$

where N_n is the number of parts produced with new tool, N_o is the number of parts produced with old tool, t_n is the time required to produce each part with new tool, t_o is the time required to produce each part with old tool, R_n is the time charge rate for new tool, R_o is the time charge rate for old tool, and C_t is the total cost of new tool.

To determine the break-even point in terms of the number of parts required in a production run to justify the expenditure in the new tool, let $R_n = R_o$ and let N be the break-even point for both applications, that is, $N = N_n = N_o$. As a result, Eq. 1.15 can be written as

$$N t_n R + C_t = N t_o R$$

from which the break-even point N can be isolated as

$$N = \frac{C_t}{R(t_o - t_n)} \tag{1.16}$$

where $t_o - t_n$ is the time savings per part as a result of using the new tool. Another approach at solving for the break-even point is to write that

Cost of new tool = Number of parts produced × time saved × charge rate
= Savings through use of tool

or

$$C_t = N(t_o - t_n)R$$

from which

$$N = \frac{C_t}{R(t_o - t_n)} \tag{1.16}$$

EXAMPLE 1.11. A new tool application can increase efficiency by saving 20 sec off the production time of each part manufactured. The cost of the tool is $375.00. If the labor and overhead are $30.00/hr, find the break-even point to justify the expenditure in the tool.

From Eq. 1.16,

$$N = \frac{375}{30(0.0055)} = 2273 \text{ parts}$$

where

$$t_o - t_n = 20 \text{ sec} = 0.33 \text{ min} = 0.0055 \text{ hr}$$

This example illustrates that unless the production exceeds 2273 parts, the application of the new tool will result in a net loss for the operation. The production must exceed the break-even point in order to generate a profit.

In the case where the decision to incorporate a new tool into an operation is not limited to a single choice, the use of a break-even chart may prove to be useful. Figure 1.7 illustrates a break-even chart for two different tools.

As can be seen from Fig. 1.7, tool 2 is more expensive than tool 1, as indicated by the cost (C_{t2} versus C_{t1}), but produces at a more efficient rate, as documented by the slope of the line. Production range r_1 defines a span where it is uneconomical to use either tool 1 or tool 2 insofar as they both will produce a loss, since the initial investments (C_{t1} and C_{t2}) are not recovered as a gain because the number of parts produced is too low to offset the cost of the tools. Break-even point n_1 defines the point at which it begins to become economical to use tool 1. Beyond this point, tool 1 generates a profit. Range r_2 defines a

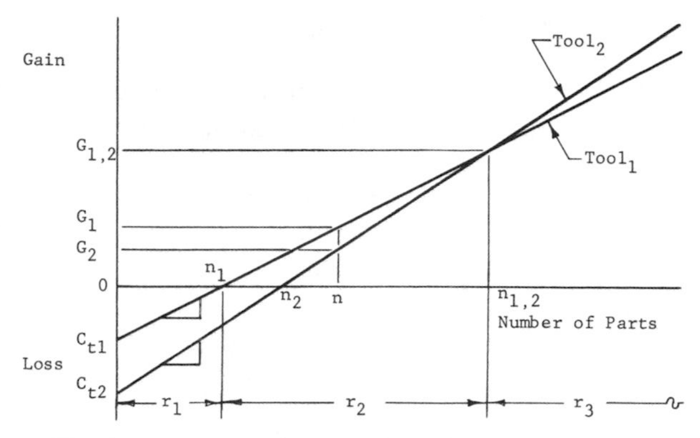

Fig. 1.7. Break-even chart for two different tools.

span of production where it is obviously more economical to use tool 1. At point n_2, tool 2 becomes economical to use but possesses a disadvantage over tool 1 because of its higher initial cost. The same gain ($G_{1,2}$) is achieved at point $n_{1,2}$, the break-even point between tools 1 and 2. Beyond point $n_{1,2}$ it is more economical to use tool 2 as described by range r_3.

The gain from use of the tool can be written as

$$G = gn - C_t \tag{1.17}$$

where G is the gain or loss obtained through the use of a tool for a given production run, g is the gain contributed by a tool per part produced, n is the number of parts produced, and C_t is the cost of the tool. In order to solve for n_1 (the break-even point for a single tool in terms of the required number of parts to be produced), G is set at zero and the equation is solved for n. As a result,

$$n = \frac{C_t}{g} \tag{1.18}$$

To solve for the break-even point between two tools ($n_{1,2}$) we let

$$G_{1,2} = G_1 = G_2$$

$$G_{1,2} = g_1 n_{1,2} - C_{t1} = g_2 n_{1,2} - C_{t2}$$

Solving for $n_{1,2}$ yields

$$n_{1,2} = \frac{C_{t2} - C_{t1}}{g_2 - g_1} \tag{1.19}$$

Equation 1.19 can be expressed as follows:

$$\begin{array}{c}\text{Break-even point}\\ \text{for two tools}\end{array} = \frac{\text{Difference in cost of tools}}{\text{Difference in gain per part}}$$

EXAMPLE 1.12. Two tools are compared for profitability. Tool 1 costs \$100.00 and enables a profit margin of \$0.20 per part to be maintained on a production run, whereas tool 2 costs \$200.00 and enables a profit margin of \$0.25 per part. Find the production ranges for which each of these tools should be recommended.

From Eq. 1.18, n_1 and n_2 can be evaluated as follows:

$$n_1 = \frac{100}{0.2} = 500 \text{ parts}$$

$$n_2 = \frac{200}{0.25} = 800 \text{ parts}$$

The break-even point $n_{1,2}$ can be evaluated from Eq. 1.19 as

$$n_{1,2} = \frac{200 - 100}{0.25 - 0.20} = 2000 \text{ parts}$$

From the data listed, it can be concluded that the operation is unprofitable below 500 parts insofar as the investment in tool 1 is not recovered by the number of parts produced. Between a production run of 500 and 800 parts, tool 1 should be used since it yields a gain. Beyond a run of 800 parts, tool 2 should be used since it possesses an advantage over tool 1.

To determine the advantage of one tool over the other, it can be written that

$$G_2 - G_1 = n(g_2 - g_1) - (C_{t2} - C_{t1}) \tag{1.20}$$

For the example listed above, a production run of 4000 parts would yield

$$G_2 - G_1 = 4000(0.05) - 100$$

$$G_2 - G_1 = \$100.00 \text{ advantage of tool 2 over tool 1}$$

For a production run of 1000 parts, the results would be

$$G_2 - G_1 = \$50.00 \text{ disadvantage of tool 2 over tool 1}$$

A graphical display of the data in example 1.12 is given in Fig. 1.8.

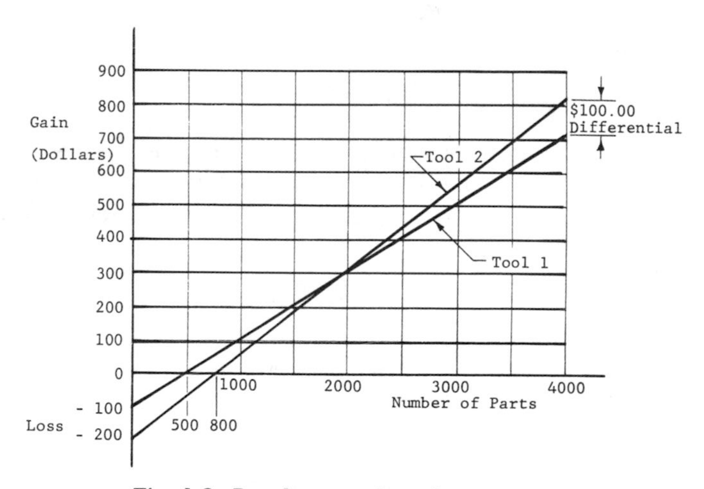

Fig. 1.8. Break-even chart for two tools.

EXAMPLE 1.13. Two new tool types are being considered for a series of production applications.

Tool	Cost ($)	Anticipated Gain per Part Produced ($)
1	600	0.50
2	1800	0.75

For a production run of 3000 parts, which tool would you recommend? How much can be anticipated as a total gain from the use of the tool that has the greatest potential?

From Eq. 1.20,

$$\begin{aligned} G_2 - G_1 &= 3000(0.25) - 400 \\ &= \$350.00 \text{ gain over tool 1} \end{aligned}$$

Solving for the total gain from the use of tool 2 yields, from Eq. 1.17,

$$\begin{aligned} G_2 &= 0.75(3000) - 1000 \\ &= \$1250 \end{aligned}$$

Another approach in appraising the impact of a more efficient tool is to evaluate the tool investment limit that can be expended on a fixed production run. If the tool expenditure has to be justified in terms of its direct profit contributions within a number of parts produced, then Eq. 1.16 can be solved for corresponding tool cost. Then

$$C_t = nR(t_o - t_n) \qquad \textbf{(1.21)}$$

Equation 1.21 is given in terms of time savings through the use of the new tool over the old tool. From a direct dollar gain, the tool investment limit can be determined from Eq. 1.18:

$$C_t = gn \qquad \textbf{(1.22)}$$

EXAMPLE 1.14. What is the tool investment limit that can be expended for a more efficient tool that has the potential of saving 20 sec/part in an operation that has a charge rate of $30.00/hr for a production run of 5000 parts?

From Eq. 1.21, where

$$\begin{aligned} R(t_0 - t_n) &= g \\ &= 30(20)(1/60)(1/60) \\ &= 0.1667 \text{ dollars/part} \end{aligned}$$

we obtain

$$C_t = 5000(0.1667) = \$833.50$$

1.10 TOOL COST AND SIZE OF PRODUCTION

The size of production has a profound effect not only on the amount that can be allocated for tooling but also on the tooling cost that is charged against each part produced. The case where the entire cost of a tool must be absorbed in a given production run will be considered in this section.

Usually, the cost of machining a number of parts can be expressed in terms of the tooling cost plus a fixed cost that can be assigned to each individual part. If the production run is low and the tooling cost must be charged to the run, the effect of the tooling cost can have a significant impact on the unit cost. On the other hand, high production runs can reduce the significance of the tooling cost per unit by dividing this cost over a large number of parts.

A simplified approach to writing the cost of production in terms of the tooling cost plus other costs associated with manufacturing is as follows:

$$\text{Cost of production} = \text{Tooling cost} + \text{fixed cost}$$

For the convenience of isolating the tooling cost, all other costs are combined under the label of fixed cost. Algebraically this can be expressed as

$$C_p N = C_t + C_f N \tag{1.23}$$

where C_p is the cost of production per part, N is the number of parts produced, C_t is the total tooling cost, and C_f is the fixed cost per part. Since the tooling cost as well as the fixed cost would usually be known, it is convenient to write Eq. 1.23 in terms of the production cost per part as

$$C_p = \frac{C_t + C_f N}{N} \tag{1.24}$$

To dramatize the influence of the size of a production run on the distribution of tooling cost for the number of parts produced, the following example is given. This example illustrates how the cost of production varies as a function of the number of parts produced for different production runs where the tooling cost is $300.00 and the fixed cost per part is $3.00 (Table 1.6). A plot of the cost per part versus the size of

Table 1.6. Cost per Part and Production Size

N:	10	100	1000	2000	3000	4000	5000
C_p ($):	33.00	6.00	3.30	3.15	3.10	3.08	3.06

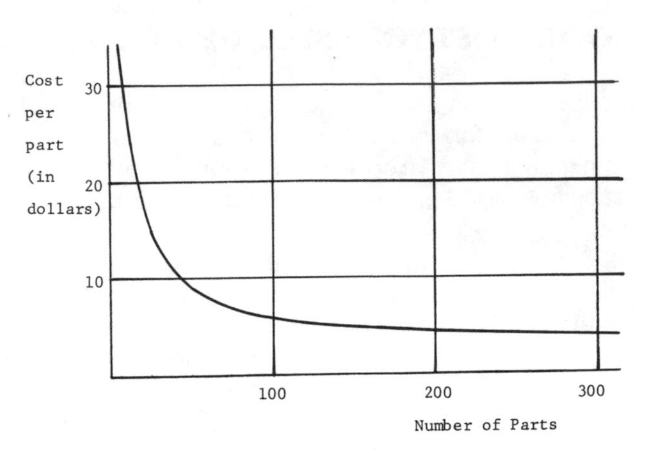

Fig. 1.9. Manufacturing cost per part versus production run.

the production run is illustrated in Fig. 1.9. The results listed in Table 1.6 illustrate the significance of the tooling cost on the cost per part. Increasing the run from 10 to 1000 parts reduces the cost per part by a factor of 10. At a production run of 5000 parts, the tooling cost of $0.06 represents 2% of the cost per part. If it is desired to keep the tooling cost at a certain percent of the total cost, the production required to fulfill this condition can be determined by solving the cost equation for the number of parts in the production run. Equation 1.23 can then be written as

$$N = \frac{C_t}{C_p - C_f} \tag{1.25}$$

EXAMPLE 1.15. For the case illustrated in Table 1.6, what is the required production that reduces the tooling cost to 10% of the fixed cost per part?

From Eq. 1.23, where $C_t = 0.1(C_f N)$, we obtain

$$C_p N = 0.1(C_f N) + C_f N$$

from which

$$C_p = 1.1 C_f$$

Substituting into Eq. 1.25 yields

$$N = \frac{300}{3.30 - 3.00} = 1000 \text{ parts}$$

An examination of Fig. 1.9 indicates that the tooling cost has a major influence on the cost per part at low production runs. The sensitivity or slope of the cost curve is very high for low numbers of parts produced. The reason is that each part produced must carry a portion of the total cost of the tool. If the number of parts produced is low, the cost per part is high, and if the number of parts produced is high, the tooling cost per part is low because the tooling cost is spread over a large number of parts. Should it so happen that a restriction relative to the cost per part is placed on the production run, the tooling cost restriction can be calculated from Eq. 1.23 by isolating the tooling cost as

$$C_t = N(C_p - C_f) \quad (1.26)$$

EXAMPLE 1.16. Determine the amount that can be allocated for tooling in a production run of 500 parts where the fixed cost per part is set at $2.10 and the cost of production per part must be limited to no more than $2.25.

Substituting into Eq. 1.26 yields

$$C_t = 500(2.25 - 2.10) = \$75.00$$

1.11 TOOLING INVESTMENT STRATEGY FOR POTENTIAL REPEAT ORDERS

The size of the production run obviously has a profound effect on how much can be allowed to be spent for tooling. In this section, attention is brought to the case where a decision is required on an expenditure for tooling for a specific order that has a likelihood of being repeated. In many cases, when a manufacturer receives an order to produce a certain number of parts, it is for a specifically quoted total price.

The decision of how much to spend for tooling can be a crucial one since it can drastically affect the profit to be generated by the order. This is especially true in cases where there is a strong relationship between tooling cost and so-called production efficiency. For the sake of developing a quantitative approach to the tooling allocation decision, let us state the following premise: A more expensive tool is more cost effective if it can produce parts in a manner that reduces other costs of manufacturing. A typical case would be where a more expensive tool operates at a higher speed, therefore producing more parts per unit time. In this way, the cost per part of the operation is

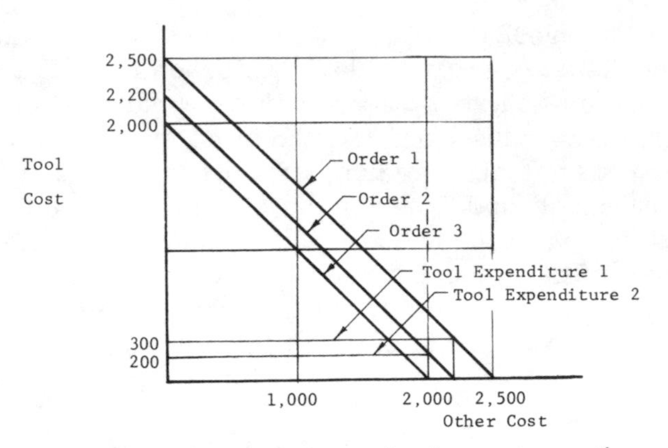

Fig. 1.10. Manufacturing cost curves for three consecutive orders.

reduced. The manufacturing cost can be written as

$$\text{Manufacturing cost} = \text{Tooling cost} + \text{other costs}$$

or

$$C_m = C_t + C_o \tag{1.27}$$

Equation 1.27 lists the manufacturing cost as being balanced between the tooling cost and other costs associated with the operation. If the manufacturing cost is fixed, as in a cost appraisal of an order, then a decision of how much to spend for tooling can affect the availability of funds for the so-called other costs.

Figure 1.10 illustrates the possible combinations between tooling cost and other costs that can be used for an order of 500 parts which has imposed upon it the manufacturing cost limit of $2500. In order number 1, $300.00 was allocated for tooling and in order number 2, an additional $200.00 was allocated for tooling. The assumption made here is that there is a direct relationship between tooling cost and other costs, that is, whatever is spent on tooling is reflected in an exact reduction in other costs. Table 1.7 lists the results of the consecutive orders on the cost per part.

Table 1.7. Results of Consecutive Orders

Order Number	Manufacturing Cost per Order ($)	Tool Cost ($)	Other Cost ($)	Cost per Part per Order ($)	Number of Parts per Order
1	2500	300	2200	5.00	500
2	2200	200	2000	4.50	500
3	2000	0	2000	4.00	500

Figure 1.10 describes the possible combinations between tooling cost and other costs that can be used for an order which has imposed upon it the manufacturing cost limit of $2500. The application used here assumes that the tooling cost is set at $300.00. The additional assumption is made that there is direct elasticity between the tooling cost and the other costs, that is, an expenditure for tooling reduces the other costs by the amount of the tooling expenditure. The other costs affiliated with the manufacturing of the 500 parts must then be limited to $2200 if the total cost of $2500 is to be maintained. In the event of a repeat order, the parts can now be manufactured for $2200 because the tooling is available from the first order. At this point, a decision for another expenditure to improve tooling can be made. In the event that an additional expenditure in tooling would yield productivity gains that would absorb the cost of the tooling, the decision to make another expenditure for tooling would be a sound one. If this were the case and an expenditure of $200.00 were made, then the availability of funds for the other manufacturing cost for the repeat order would be reduced to $2000. Again, the assumption is made that a tooling expenditure will be reflected directly in a reduction in other costs.

As the technique of improving the manufacturing process by modification of tooling continues, that is, as additional expenditures for tooling are made, the investment is converted directly to a gain with each succeeding order. The example listed in Fig. 1.9 characterizes this possibility. It can be seen that order 1 has a manufacturing cost per part of $5.00, whereas order 2 provides a means of reducing the manufacturing cost per part to $4.50. In order 3 the manufacturing cost is further reduced to $4.00/part. This reflects the direct conversion of tool investment into cost reduction.

Unfortunately, not all machining situations are as accommodating in terms of tool expenditure response to productivity gains as the case just cited. Nevertheless, the truism that there is always room for productivity advances does prevail in every single case. As the process of upgrading the tooling is repeated, the elasticity toward production improvements has a tendency to stiffen, that is, it becomes more difficult to extract these improvements. This is a result of the reduction in the funds available for each succeeding order as the improvements from tooling are extracted.

EXAMPLE 1.17. A machining appraisal totaling $20,000 is examined and indicates that a $500.00 additional tooling improvement expenditure has a direct influence on reducing other costs of the operation by $500.00. Assume that the expenditure is made and that three additional orders are received. What cost benefit would the tooling investment have?

Order	Manufacturing Cost ($)	Tooling Expenditure	Other Cost	Cost Benefit per Order	Total Cost Benefit ($)
1	20,000	$500	19,500	0	0
2	19,500	0	19,500	500	500
3	19,500	0	19,500	500	1000
4	19,500	0	19,500	500	1500

EXAMPLE 1.18. What would be the cost benefit in example 1.16 if each succeeding order involved a $500.00 tooling expenditure that directly influenced a reduction of other costs of manufacturing?

Order	Manufacturing Cost ($)	Tooling Expenditure ($)	Other Cost ($)	Cost Benefit Per Order ($)	Total Cost Benefit ($)
1	20,000	500	19,500	0	0
2	19,500	500	19,000	500	500
3	19,000	500	18,500	1000	1500
4	18,500	500	18,000	1500	3000

In examining examples 1.16 and 1.17, it becomes evident that an expenditure in tooling gains cost benefits with succeeding orders. A continuous upgrading of the operation as indicated by example 1.17 obviously has an advantage insofar as each succeeding order contributes a larger amount toward the cost benefit when compared with the original order. In the case where there is an uncertainty that an order will be repeated, the tooling expenditure decision is left to speculation. If a direct relationship between the tooling expenditure and the reduction in other costs exists, then there is no loss in making the investment in new tooling. On the other hand, if the order were to be repeated, then the tooling expenditure would be converted into a cost reduction advantage.

1.12 A SUMMARY OF ECONOMIC CONSIDERATIONS

The examples given in this chapter represent an attempt to illustrate in a general fashion how an economic analysis can lead to a more efficient and less costly machining operation. The topic dealing with the proper economic use of cutting tools can be a complex one. What may seem to be a suitable use under one set of circumstances may not be practical under different conditions. Take the case of the best setting of cutting speed. A response curve of production versus cutting speed similar to Fig. 1.1 can be determined by gathering experimental data through measurement of the production rate as the cutting speed is varied. To determine whether an operation is functioning above or below the idealized cutting speed, in an attempt to determine the one that produces the maximum production rate, all that needs to be done is to change the cutting speed and test whether the production rate has increased or decreased. If an increase in cutting speed results in increased returns, then the operation is below the maximum production rate. Should it so happen that the opposite is true, that is, increases in cutting speed result in diminishing returns, then the operation is above the maximum production rate. In a similar fashion, the maximum production rate can be determined when the cutting speed is set at maximum production by changing the existing cutting speed. If either an increase or decrease in cutting speed results in a decrease in the production rate, then the conclusion can be drawn that the operation was set at maximum production.

The question of whether or not a cutting speed setting at maximum production is the proper one depends on many factors. In the case where delivery schedules must be met regardless of cost, then this cutting speed setting will produce the maximum number of parts per unit time. On the other hand, should it so be that importance is placed on maintaining profit margins, then the cutting speed setting at maximum production may not be the correct one insofar as the operation may be an expensive one. To test this condition, data could be collected on manufacturing cost as a function of cutting speed, and these data could be plotted to produce a graph similar to Fig. 1.3. One would know that the cutting speed setting is at minimum cost if any change from this setting generates an increase in the cost of the operation. A comparison of the data gathered on production and cost as a function of cutting speed could then be used to balance the trade-off between minimum cost and maximum production operations.

A more efficient tool in many cases can be defined as one that functions in such a fashion as to produce more parts in less time. A more efficient tool may also possess characteristics, such as more convenient positioning and tool changing, that contribute toward reducing the downtime of an operation. The production rate increase of a more efficient tool can be determined by comparing its production rate versus the production rates of other tools.

To justify the additional cost of a more efficient tool, an analysis describing break-even points can be used. When comparing different tools, break-even points in terms of numbers of parts produced can be examined to describe ranges of production for which certain tools should be recommended since they make a greater contribution toward profitability.

The size of the production run has a significant effect on the manufacturing cost per part because each part produced absorbs a portion of the tooling cost. Large production runs can reduce the manufacturing cost to values that are close to the cost that is fixed on a part, such as material cost and other fixed expenses. On the other hand, low production runs can result in very high manufacturing costs, attributable directly to writing off the cost of tooling against a small number of parts produced.

Repeat orders provide an advantage for a manufacturer to capitalize on a tooling expenditure provided that available tooling from a previous order can be used again. Should this be the case, manufacturing costs for the repeat order are reduced to reflect the benefit of available tooling. In some cases, this provides an opportunity to analyze an operation to determine if additional tooling expenditures are justifiable.

PROBLEMS

PROBLEM 1.1. The following data have been compiled from a series of tests on production rate.

Production rate (parts/hr):	10	18	19.5	14	8
Cutting speed (ft/min):	150	175	215	240	255

Plot the points on graph paper and draw a smooth curve through these points. From the graph determine: (a) the maximum production rate and the corresponding cutting speed and (b) the two cutting speeds that would yield a production rate of 12 parts/hr.

Answers: (a) $P_{rm} = 20$ parts/hr
$V_m = 200$ ft/min
(b) $V_1 = 155$ ft/min
$V_2 = 245$ ft/min

PROBLEM 1.2. Determine the downtime per hour that is associated with a production rate of 12 parts/hr at a cutting speed of 47.24 m/min. A test reveals that 12 parts/hr can also be produced with a cutting speed of 74.68 m/min which results in a downtime of 25 min/hr. Assume that the machine rate is proportional to the cutting speed.

Answer: $T_{d1} = 4.68$ min/hr

PROBLEM 1.3. What production rate can be expected from an operation where the machine rate is 0.1333 parts/min and the machine downtime is 12 min/hr?

Answer: $P_r = 6.4$ parts/hr

PROBLEM 1.4. The following data have been compiled from a series of evaluations of costs on a given operation.

Cost ($/part):	4.50	4.20	4.04	4.02	4.10	4.60
Cutting speed (m/min):	38.1	42.7	48.8	56.4	61.0	70.1

Plot the points on graph paper and draw a smooth curve through these points. From the graph, determine: (a) the minimum cost for this operation and the corresponding cutting speed and (b) the two cutting speeds that would result in the same cost per part of $4.30.

Answers: $C_n = \$4.00$/part
$V_n = 53.34$ m/min
$V_1 = 41.15$ m/min
$V_2 = 65.23$ m/min

PROBLEM 1.5. Two cutting speeds yield a production rate of 25 parts/hr. While the machine is operating, the first cutting speed setting produces parts at a rate of 27.27 parts/hr, and the second cutting speed setting produces parts at a rate of 36.36 parts/hr. A tool cost analysis indicates that for the first cutting speed setting, the tool cost per part is $0.25, whereas for the second cutting speed setting it is $0.50/part. If the machine charge rate is $35.00/hr, determine the downtime per hour and the cost per part for the first and for the second settings of cutting speed.

Answers: $R_{d1} = 5$ min
$T_{d2} = \$18.75$/min
$C_{v1} = \$1.65$/part
$C_{v2} = \$1.90$/part

PROBLEM 1.6. A cost analysis indicates that a cutting speed setting yields a cost of manufacturing of $3.80/part. The first setting has associated with it a machining cost of $2.90/part, a maintenance downtime cost of $0.60, and a tool depreciation and wear cost of $0.30/part. It is noted that a 15% increase in cutting speed reduces the tool life by 30%. Determine the cost per part for the higher setting of the cutting speed and compare the production rate for the higher and lower settings of the cutting speed.

Answers: $C_{v2} = \$3.80$/part
$\Delta P_r = 3.6\%$

PROBLEM 1.7. Two cutting speed settings cause the cost of manufacturing for a given operation to be $4.30/part. The lower cutting speed has affiliated with it a tool cost of $0.21/part, whereas the higher cutting speed has affiliated with it a tool cost of $0.53/part. Determine the increase in production of the second setting over the first setting of cutting speed.

Answer: $\Delta P_r = 8.5\%$

PROBLEM 1.8. Two different tools are tested. In two separate tests, each of 8-hr duration, tool 1 produced 27 parts, with a corresponding downtime of 56 min; tool 2 produced 31 parts, with a corresponding downtime of 45 min. Determine how long it would take to fill an order of 500 parts using each of these tools. In addition, calculate the change in production rate of tool 2 over tool 1.

Answers: $t_1 = 148$ hr
$t_2 = 129$ hr
$\Delta\% = 14.8\%$

PROBLEM 1.9. Find the production rate, while the machine is running (P_2/t_2), of an experimental tool that yields a 25% increase in the production rate over an existing tool that is in use. The existing tool has an actual production rate of 12 parts/hour. It is found that the experimental tool has a downtime of 10% of the machining time.

Answers: $P_{a2} = 15$ parts/hr
$P_2/t_2 = 16.5$ parts/hr

PROBLEM 1.10. Find the break-even point (number of parts) in the use of a new tool over an older one in the case where the expenditure for the new tool is $200 and the additional contribution toward profitability of the new tool is $0.32/part. In addition, determine how much can be gained by using the new tool for a production run of 2500 parts.

Answers: $n = 625$ parts
$G = \$600$

PROBLEM 1.11. Two tools are considered for use in a production run of 300 parts. Tool 1 has an initial cost of $110 and enables the parts to be produced at a gain of $0.55 each. Tool 2 has an initial cost of $218, runs more efficiently than tool 1, and results in a gain of $0.63 for each part produced. (a) Which tool would you recommend for use? (b) How much can be earned by running this order?

Answers: (a) $n_{1,2} = 1350$ parts; since $n_{1,2} > 1300$, use tool 1.
(b) $G_1 = \$605$
$G_2 = \$601$

PROBLEM 1.12. What is the monetary advantage of tool 2 over tool 1 for the cases listed below?

Case 1: Production run 100 parts
Case 2: Production run 375 parts
Case 3: Production run 2000 parts
Tool 1: Gain contributed = \$0.25/part; cost of tool 1 = \$85
Tool 2: Gain contributed = \$0.32/part; cost of tool 2 = \$133

Answers:

$$n_1 = 340 \text{ parts}$$
$$n_2 = 416 \text{ parts}$$
$$n_{1,2} = 686 \text{ parts}$$

Case 1: Since $100 < n_1 < n_2$, both tools produce a loss:

$$G_2 - G_1 = -\$41$$

Case 2: Since $n_1 < 375 < n_2$, use tool 1:

$$G_2 - G_1 = -\$21.75$$

Case 3: Since $n_2 < 2000$, use tool 2:

$$G_2 - G_1 = \$92$$

PROBLEM 1.13. Two tools are considered for use. Tool 1 costs \$125 and enables a savings of \$0.30/part to be made. Tool 2 costs \$175 and enables a savings of \$0.38/part to be made. Determine the break-even point between the two tools and the largest gain that can be expected from a production run of 2100 parts.

Answers: $n_{1,2} = 625$ parts
$G_2 = \$623$

PROBLEM 1.14. It is desired to reduce the cost per part to \$5.78 in the case where the tooling cost is \$475 and the fixed cost per part is \$5.15. How large must the production run be in order to satisfy this case?

Answers: $N_p = 754$ parts

PROBLEM 1.15. How much can be allotted for tooling in the case where the fixed cost per part is \$3.15, the cost per part cannot exceed \$3.40, and the production run is 1200 parts?

Answer: $C_t = \$300$

PROBLEM 1.16. What percentage of the cost per part is the tooling cost for the case where the fixed cost per part is \$2.85, the tooling cost is \$375, and the production run is 975 parts?

Answer: $\%\ C_{pt} = 11.9\%$

PROBLEM 1.17. A given tool produces parts at a rate of 150 parts in an 8-hr period, with a corresponding downtime of 15% of the machine time. It is proposed to use a more efficient tool that can produce 180 parts in an 8-hr period and can yield a corresponding downtime of 10% of the machine time. Determine the production percentage increase that can be gained by using the more efficient tool.

Answer: $\Delta\% = 25.46\%$ increase

PROBLEM 1.18. Five hundred parts are to be manufactured at a cost of \$6250. Determine: (a) how much can be allocated for tooling costs if all other costs are calculated to be \$9.87/part; and (b) the amount for which a repeat order can be completed if the tool is available from the first order.

Answers: (a) $C_t = \$1315$
(b) $C_m = \$4935$

PROBLEM 1.19. If 10% of manufacturing costs are set aside for tooling that upgrades a manufacturing process, what percentage of the first order can a fourth repeat order be manufactured for? Assume that the tooling cost is converted directly to a manufacturing cost reduction on consecutive repeat orders.

Answer: $\% = 72.9\%$

PROBLEM 1.20. A cost per part quotation is given for an order of 100 parts that has affiliated with it a tooling cost of \$580 and a fixed cost per part of \$3.94. It is desired to reduce the cost per part by a factor of 40%. What is the required production size to achieve this goal?

Answer: $N_{p2} = 305$ parts

PROBLEM 1.21. Determine the two cutting speeds that will result in a production rate of 90 parts/hr if a manufacturing operation is described by the equation listed below:

$$P = -2.5 \times 10^{-3} V^2 + V$$

Answers: $V = 136.8,\ 263.2$ ft/min

PROBLEM 1.22. What cutting speed would you recommend for maximum production for the operation described in problem 1.21?

Answer: $V_m = 60.96$ m/min

PROBLEM 1.23. The following results were attained from a series of tests analyzing costs of production as a function of the cutting speed.

Test	V	C_p
1	150	3.37
2	200	3.17
3	250	5.04

From the given data, derive an expression for the cost of production as a function of cutting speed.

Answer: $C_p = 4.151 \times 10^{-4} V^2 - 0.1492V + 16.443$

PROBLEM 1.24. What cutting speed would you recommend for minimum cost operations for the machining condition described in problem 1.23?

Answer: $V = 55.47$ m/min

PROBLEM 1.25. A tooling expenditure is $500.00 for an operation that has a fixed cost per part of $2.00. Determine the required production that will make the tooling expenditure 15% of the fixed cost per part.

Answer: $N = 1667$ parts

BIBLIOGRAPHY

ASME Research Committee on Metal Cutting Data and Bibliography, *Manual on Cutting of Metals*, American Society of Mechanical Engineers, New York, 1952.

Dallas, Daniel B., *Tool and Manufacturing Engineers Handbook*, Society of Manufacturing Engineers, Dearborn, Michigan, 1976.

Sedlik, Harold, *Jigs and Fixtures for Limited Production*, Society of Manufacturing Engineers, Dearborn, Michigan, 1970.

Steffy, Wilbert, *Economics of Machine Tool Procurement*, Society of Manufacturing Engineers, Dearborn, Michigan, 1978.

2 The Cutting Process

2.1 INTRODUCTION

The technique by which metal is cut is a complex one. It involves a specially shaped tool exerting a concentrated force on the work material. The effects of this force are distributed in such a fashion as to cause a controlled plastic flow that results in the formation of a scrap chip. The objective of the process is to remove excess material from an oversized workpiece. In this way, the workpiece is shaped to conform with dimensional and surface finish specifications.

There are four components that interact in all metal cutting operations: the tool, the workpiece being machined, the work holder, and the tool holder. Usually, the tool and workpiece are rigidly held in their respective holders, which are mounted on the machine tool. In order for cutting to take place, the tool must be guided at a relative velocity, cutting speed, toward the workpiece. The control over how the workpiece is to be machined is determined through two different machine settings. One is referred to as *the feed*, the rate at which the tool advances, whereas the other is referred to as *the depth of cut*, the amount to be removed with one pass of the tool.

In order for the tool to perform its function, it must possess certain characteristics. Among these is the tool's hardness, which must be greater than that of the material being cut. This enables the tool to absorb the forces of the cutting process without distortion. The shape of the tool is also important. It usually has a wedgelike shape for effective penetration into the workpiece. Wear resistance and retention of hardness at elevated temperatures are also critical characteristics of a tool. They provide for durability. This, in turn, allows the tool to endure the harsh conditions under which it must function.

In this chapter, the single-point tool is introduced. The control of the tool through settings of feed, depth of cut, and cutting speed is also illustrated in the context of various machining operations. The cutting process is then diagnosed to develop an appreciation of the phases through which an element of the workpiece moves in the process of forming the finished shape of the workpiece.

The standard stress–strain experiment is elaborated on in this chapter to point out the physical similarities between a test specimen and a machined workpiece. This leads to the introduction and application of common engineering terms such as work and thermal energy. It is shown that these are related through the first law of thermodynamics. Documentation is provided by the famous experiment of James P. Joule. The developed concepts are then applied in calculating the temperature rise of a chip in the metal cutting process. Further analysis involves the development of a simplified model of the cutting process.

Toward the end of this chapter, an analytical derivation of an expression to determine the shear angle in the cutting process is given. This is further expanded upon by a description of an experiment dealing with the effect of the tool rake angle on the shear angle of the metal cutting process. Finally, the classification of different types of chips, by-products of the metal cutting process, is described.

2.2 SINGLE-POINT TOOLS

Metal cutting operations involve a tool that removes excess material from a workpiece. Thus, the workpiece is shaped to conform with a given dimensional specification. Figure 2.1 gives an example of a typical single-point tool.

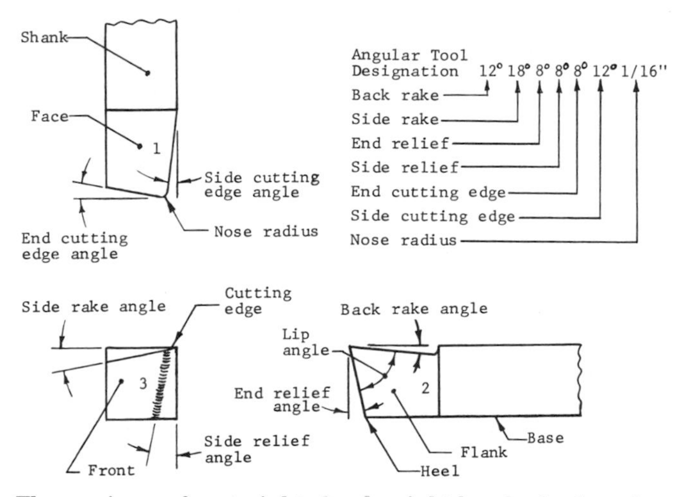

Fig. 2.1. Three views of a straight shank, right-hand, single-point tool, with an example of angular tool designation.

Of interest is the description of the tool that takes into account the three surfaces (1, 2, and 3). Each of these surfaces is defined by two angles relative to the shank of the tool. These angles are ground into the tool in order to form the cutting edge. If one were to experimentally attempt to cut metal with a tool, it would be discovered that an effective tool would have a shape similar to that shown in Fig. 2.1. It is shaped in a wedgelike fashion that enhances cutting. Clearance angles are provided to enable cutting to take place without any rubbing action.

As can be seen in Fig. 2.1, the tool face, surface 1, is formed by grinding in the side rake angle and the back rake angle. The tool flank, surface 2, is formed by grinding in simultaneously the side relief angle and the side cutting edge angle. The front of the tool, surface 3, is formed by grinding in the end relief angle and the end cutting edge angle. Finally, the nose radius is ground into the tool to provide for a smooth machined surface on the workpiece. The purpose of the nose radius is to eliminate the threadlike groove that would be produced by a sharply pointed tool. Without a nose radius, a threadlike groove would be formed in the workpiece in a winding helical shape as a result of the relative motion between the tool and the workpiece.

The application of a single-point tool is shown in Fig. 2.2 in a turning operation. As can be seen, the workpiece is rotating with an angular velocity, ω, usually given in terms of revolutions per minute (rpm). The velocity of the workpiece relative to the cutting tool is referred to as the *cutting velocity* or *cutting speed*. Control over how much material is to be removed from the workpiece is determined through machine settings of the feed, given in inches per revolution, as well as the depth of cut, given in units of inches. Figure 2.3 illustrates how the excess material is removed from the workpiece in the form of the

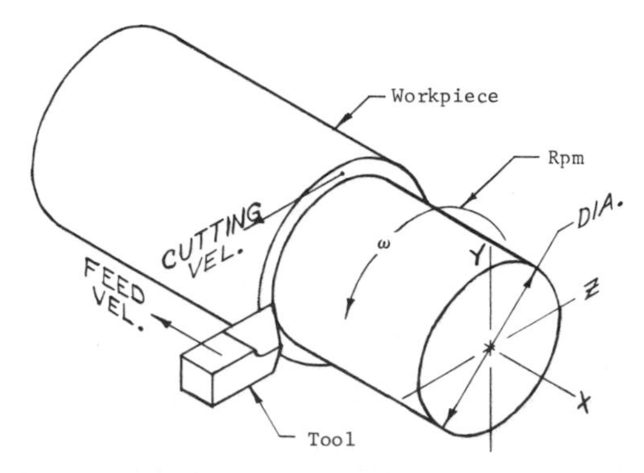

Fig. 2.2. Tool–workpiece relationship for a turning operation.

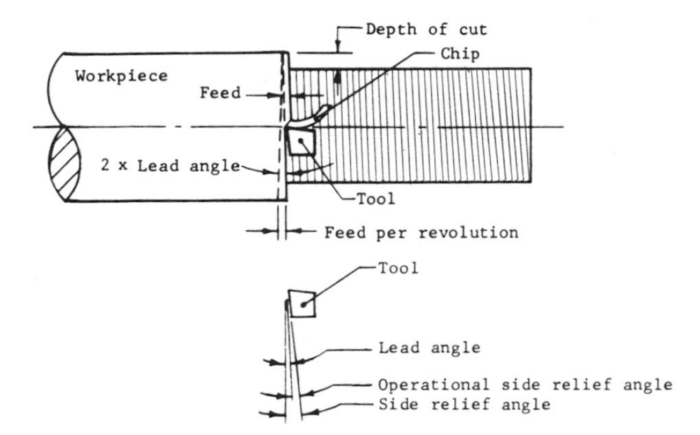

Fig. 2.3. Turning operation.

scrap chip. Figure 2.4 shows a side view of the turning operation illustrating the cutting speed, a machine setting.

A by-product of the turning operation is the generation of scrap in the form of a ribbonlike string, called a *chip*. A measure of the volume occupied by these chips can be represented in terms of the machine settings. Figure 2.5 represents a model of the machine settings for a turning operation.

The machining operation known as *boring* is one that is similar to turning. The difference is that boring generates internal shapes whereas turning generates external shapes. Both operations generate a variety of forms. These include continuous diameters, a series of stepped diameters, as well as tapered sections and contoured surfaces. Figure 2.6 illustrates a single-point tool mounted on a boring bar and gives a full-section diagram of a tapered and constant-diameter workpiece. The functions of cutting speed, feed, and depth of cut are the same as for turning, except they are internal rather than external.

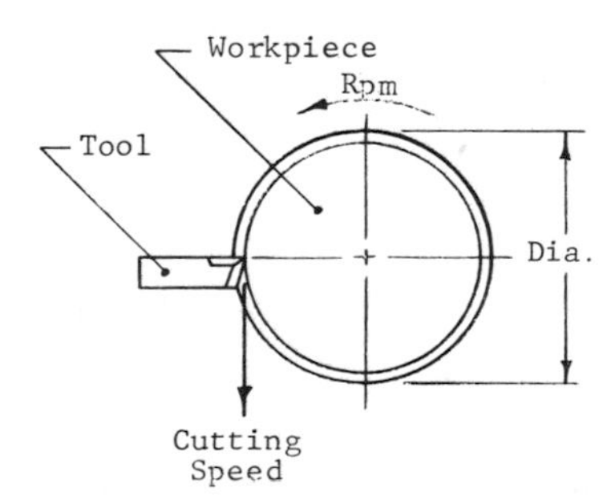

Fig. 2.4. Side view of turning operation.

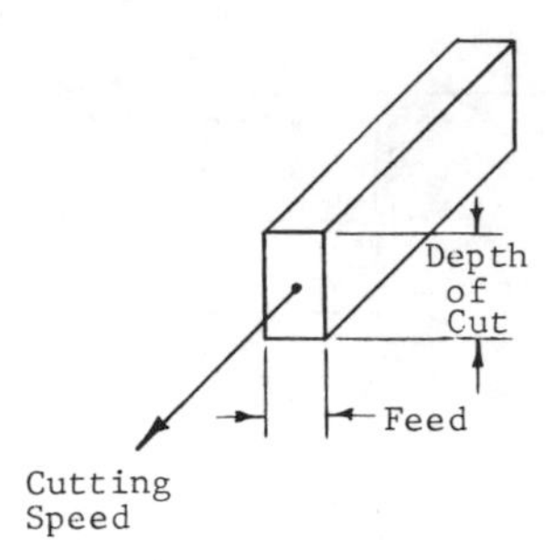

Fig. 2.5. Model of volumetric rate of machining.

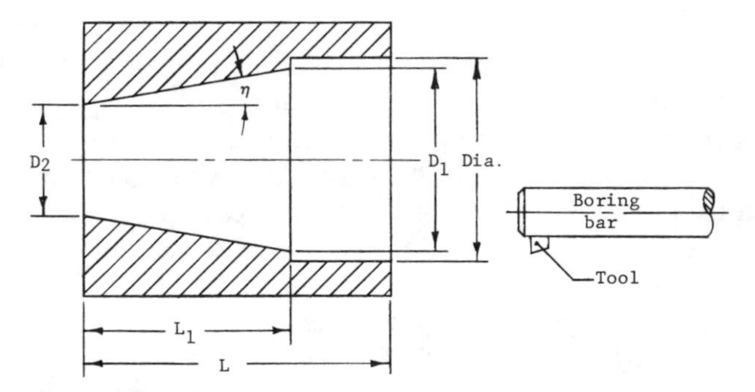

Fig. 2.6. Full-section top view of a boring operation.

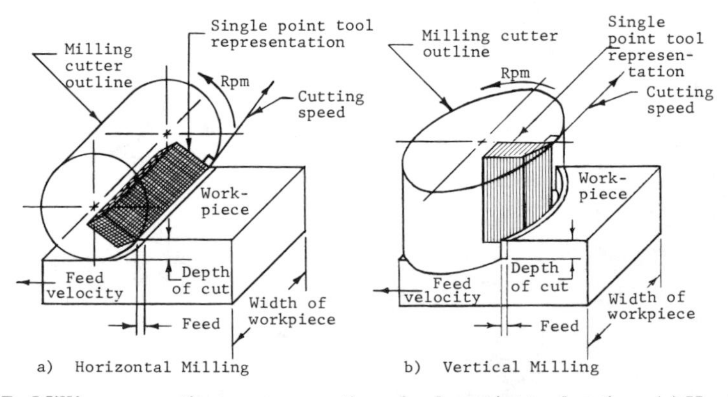

Fig. 2.7. Milling operations representing single-point tool action. (a) Horizontal milling. (b) Vertical milling.

Illustrated in Fig. 2.7 is the milling operation. It is a machining process that utilizes a multiple-edged cutter to generate surfaces by removing a certain amount of material from an oversized workpiece. The diagram is simplified to show the milling cutter reduced to a single-point tool with the machine settings of cutting speed, feed, and depth of cut. The amount of material removed per cutting edge is represented by the areas labeled feed and depth of cut.

Most metal cutting operations can be characterized by a single-point tool operating with corresponding machine settings. Figure 2.8 illustrates the shaping and drilling operations. In the shaping operation, the tool travels in a straight-line reciprocating motion and the workpiece is fed into the cutting tool. If the drilling operation is analyzed in terms of a single-point tool, then the depth of cut can be represented in terms of the radius of the drill and the feed can be represented in terms of the advance per revolution of the drill in the axial direction.

In examining the machining operations that have been illustrated in the preceding diagrams, it can be seen that the cutting speed is a function of the rpm setting. In most cases, a recommended cutting speed for an operation involving a tool of a given type in affiliation with a specific workpiece material is specified in the literature. A calculation then has to be performed to determine the rpm setting. For the turning example given in Fig. 2.4, it can be written that

$$\text{Rpm} = \frac{12V}{\pi D} \tag{2.1}$$

where V is the cutting speed (ft/min), $\pi = 3.1416$ (a constant), D is the diameter (in.), and 12 is the conversion factor (ft to in.). An inspection

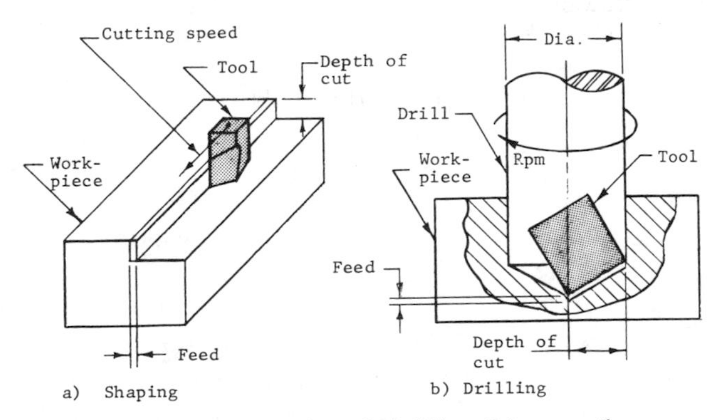

Fig. 2.8. Shaping (a) and drilling (b) operations.

of the terms in Eq. 2.1 indicates that the equation can be expressed in the form

$$\text{Rpm} = \frac{\text{Cutting speed}}{\text{Circumference or distance traveled per revolution}}$$

EXAMPLE 2.1. Calculate the rpm machine setting in order to turn a 2-in. (50.8-mm)-diameter gray cast iron workpiece with a throw-away carbide insert tool. The feed is 0.015 in. (0.381 mm) per revolution (rev), the depth of cut is 0.150 in. (3.81 mm), and the recommended cutting speed is 200 ft/min (60.96 m/min).

Substituting into Eq. 2.1 yields

$$\text{Rpm} = \frac{12\text{ in.}}{\text{ft}} \times \frac{200\text{ ft}}{\text{min}} \times \frac{1}{3.14} \times \frac{\text{rev}}{2\text{ in.}}$$

$$= 382.2\text{ rpm}$$

The volumetric rate of metal removal gives a measure of the amount of material that is being cut per unit of time. It is an important factor in machining because it is related to the energy and forces involved in a metal cutting operation. Figure 2.5 represents a model of the volumetric rate of machining. It is expressed in terms of the three machine settings and can be written as

$$R_v = 12Vfd \tag{2.2}$$

where R_v is the volumetric rate (in.^{3}/min), V is the cutting speed (ft/min), f is the feed (in./rev), d is the depth of cut (in.), and 12 is the conversion factor (ft to in.).

EXAMPLE 2.2. From the data given in example 2.1, evaluate the volumetric rate of machining.

Substituting into Eq. 2.2 gives

$$R_v = 12(200)(0.015)(0.150)$$

$$= 5.4\text{ in.}^3\text{/min } (1.474\text{ cm}^3\text{/sec})$$

For a milling operation where the feed velocity is set on the machine, the volumetric rate of machining can be expressed as

$$R_v = F_v dw \tag{2.3}$$

where F_v is the feed velocity (in./min), d is the depth of cut (in.), and w is the width of cut (in.). Of interest in the milling operation is the term *feed per tooth* or *feed per cutting edge*. It is a function of not only the feed velocity but also of the rpm as well as the number of cutting edges (teeth) on the cutter. This can be expressed as

$$F_t = \frac{F_v}{N \times \text{Rpm}} \tag{2.4}$$

where F_t is the feed per tooth (in./tooth), F_v is the feed velocity (in./min), N is the number of teeth on cutter, and Rpm represents revolutions per minute of cutter.

EXAMPLE 2.3. For a milling operation similar to that shown in Fig. 2.7(a), determine the rpm setting for a 4-in.(101.6-mm)-diameter high-speed steel cutter machining gray cast iron at a recommended cutting speed of 50 ft/min (15.24 m/min).

Substituting into Eq. 2.1 yields

$$\text{Rpm} = \frac{12(50)}{3.14(4)}$$

$$= 47.7 \text{ rpm}$$

EXAMPLE 2.4. Consider the case where the milling machine feed setting for example 2.3 was 2.25 in./min (57.15 mm/min) and the depth of cut was $\frac{5}{8}$ in. (15.88 mm).

(a) For a 2-in.(50.8-mm)-wide workpiece, determine the volumetric rate of machining.

(b) Calculate the feed per tooth if the cutter of example 2.3 had eight teeth.

For milling, the volumetric rate is given by Eq. 2.3:

$$R_v = \text{Feed velocity} \times \text{depth of cut} \times \text{width}$$

$$R_v = 2.25 \text{ in./min} \times 0.625 \text{ in.} \times 2 \text{ in.}$$

$$R_v = 2.8125 \text{ in.}^3\text{/min } (0.7678 \text{ cm}^3\text{/sec})$$

The feed per tooth can be calculated from Eq. 2.4. As a result,

$$F_t = \frac{2.25}{8(47.77)}$$

$$= 0.0059 \text{ in./tooth } (0.15 \text{ mm/tooth})$$

When a cost evaluation of a machining process takes place, the time to perform the cutting is of prime importance. It has a direct influence on the cost of the operation. Figure 2.9 illustrates a shaft diameter that is to be turned to size. To solve for the time required to machine the part, it can be written that

$$\text{Time to machine part} = \frac{\text{Distance tool travels}}{\text{Feed velocity}}$$

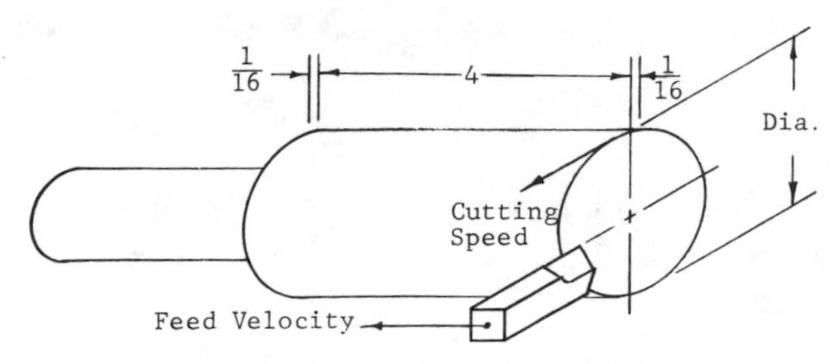

Fig. 2.9. Turning length operation.

In examining the distance the tool travels, it is noted that the tool must approach the workpiece before it begins to cut. Also, the tool must overtravel the workpiece to remove any diametral burrs. Therefore, an allowance is set aside for both approach and overtravel. In Fig. 2.9, $\frac{1}{16}$ in. (1.5875 mm) represents these allowances. Taking the allowances into account, it can be expressed that

$$\text{Distance tool travels} = \begin{array}{c}\text{Length of}\\ \text{cut}\end{array} + \begin{array}{c}\text{Approach}\\ \text{allowance}\end{array} + \begin{array}{c}\text{Overtravel}\\ \text{allowance}\end{array}$$

In a turning operation similar to that shown in Fig. 2.9, the feed velocity is not a term that is a direct machine setting. Rather, it is a product of the feed and the rpm setting of the machine tool. It is expressed as

$$\text{Feed velocity} = \text{Feed} \times \text{Rpm}$$

Taking into account these terms, the time to machine the diameter shown in Fig. 2.9 can now be written as

$$T_m = \frac{L + A_p + O_t}{f \times \text{Rpm}} \tag{2.5}$$

where T_m is the time to machine, L is the length to be machined, A_p is the approach allowance, O_t is the overtravel allowance, f is the feed/revolution, and Rpm represents revolutions per minute.

EXAMPLE 2.5. In a turning operation, a length of 4 in. (101.6 mm) is to be machined as shown in Fig. 2.9. If the feed is 0.015 in./rev (0.381 mm/rev) and the rpm setting is 382.2, determine the time required to machine the part. Assume a $\frac{1}{16}$-in. (1.5875-mm) approach and overtravel allowance.

Substituting into Eq. 2.5 yields

$$T_m = \frac{4 + 0.0625 + 0.0625}{0.015(382.2)}$$

$$= 0.72 \text{ min}$$

The approach and overtravel allowance in the milling operation can be significant because of the geometry of the cutter. Figure 2.10 illustrates the approach and overtravel relationship for a conventional milling operation. As can be seen, the approach must provide for the contour of the cutter. This offset cutter distance is labeled in the diagram by the symbol X. The Pythagorean Theorem states that for the triangle of the form shown in Fig. 2.10,

$$c^2 = a^2 + b^2$$

Applying this theorem to solve for the distance X leads to

$$\left(\frac{D}{2}\right)^2 = \left(\frac{D}{2} - d\right)^2 + X^2$$

From which

$$X = \sqrt{\left(\frac{D}{2}\right)^2 - \left(\frac{D}{2} - d\right)^2}$$

or

$$X = \sqrt{d(D - d)} \tag{2.6}$$

where d is the depth of cut (in.), D is the diameter of cutter (in.), and X is the offset cutter distance.

EXAMPLE 2.6. Determine the offset cutter distance for a 3-in.(76.2-mm)-diameter cutter that is to mill a workpiece with a depth of cut of $\frac{3}{4}$ in. (19.05 mm). The operation is illustrated in Fig. 2.10.

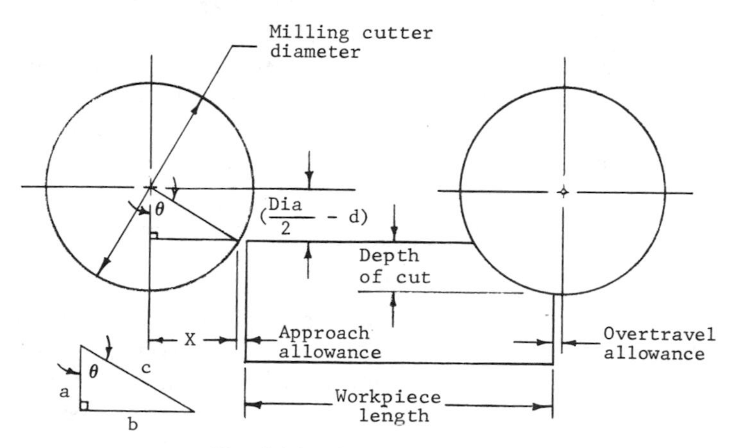

Fig. 2.10. Milling operation.

Substituting into Eq. 2.6 yields

$$\begin{aligned} X &= \sqrt{0.75(3.0 - 0.75)} \\ &= 1.299 \text{ in. } (32.99 \text{ mm}) \end{aligned}$$

EXAMPLE 2.7. Determine the time required to complete a milling operation for the case illustrated in example 2.6. The length of the workpiece is 5 in. (127 mm), the approach and overtravel allowances are $\frac{1}{16}$ in. (1.5875 mm), and the feed velocity is set at $\frac{1}{2}$ in.(12.7 mm)/min.

For the milling operation, Eq. 2.5 is altered to include the cutter offset distance and is written as

$$T_m = \frac{L + A_p + O_t + O_f}{F_v}$$

where O_f is the cutter offset distance. Substituting yields

$$\begin{aligned} T_m &= \frac{5 + 0.0625 + 0.0625 + 1.299}{0.5} \\ &= 12.848 \text{ min} \end{aligned}$$

2.3 ANALYSIS OF THE CUTTING PROCESS

During the metal cutting process, the material being cut has a concentrated force applied to it by the tool. An analysis of a single grain that is being cut from the workpiece reveals that the grain is exposed to different states of stress. Grains being cut from the workpiece move from the unstrained state up along the elastic curve beyond the yield point and into the plastic region where shear deformation takes place. Following the shear deformation, the grain is relieved of the forces acting on it and returns to an unstressed state in a deformed orientation. A closer examination of the chip-forming operation is shown in Fig. 2.11. In this example, the resultant force acting on the chip is resolved into two perpendicular components, the friction force and the normal force.

Point 1 represents a grain in the unstrained state as it approaches point 2, where the grain climbs the elastic curve to reach the plastic deformation region designated by point 3. In the plastic shear zone (3–4), the grain is stressed to a point where it plastically flows to form the chip. During the flowing process, slipping takes place and the material work hardens. Point 5 designates a point where the grain

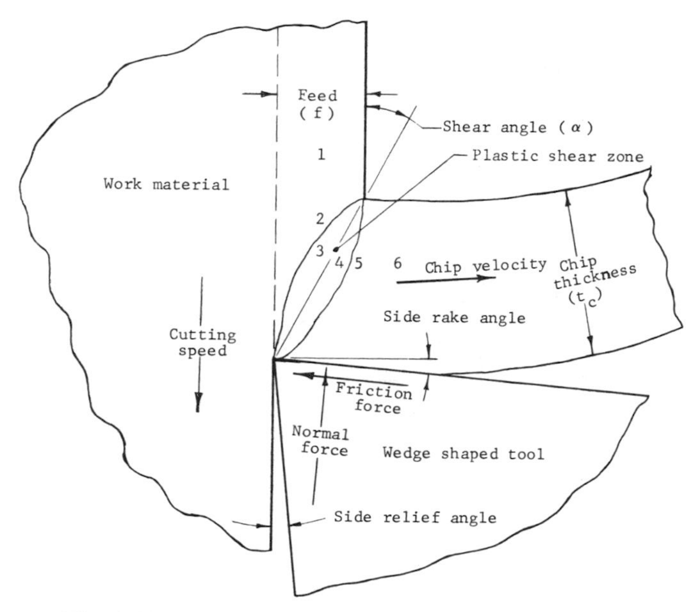

Fig. 2.11. Chip forming in metal cutting operation.

emerges from the shear zone. At this point, the grain is harder than the parent material as a result of work hardening, and it also has a higher temperature than the parent material as a result of the work performed on deforming the grain. In addition, the chip is much thicker than the feed setting from which it was generated. At point 6, the grain has been removed from the parent material and is located in the chip. It now possesses different physical properties than the parent material. Since the chip is harder, it has a higher elastic limit, that is, if the chip grain were to be cut again, a repeated point 2 would require more force for deformation. As a result, the ductility or plasticity of the chip is reduced when compared to the parent material. In addition, the chip possesses an elevated temperature from not only the plastic deformation but also from work performed on the chip by the friction force as the chip flows over the tool.

Figure 2.12 illustrates the corresponding positions of the points 1 through 6 from the chip formation diagram. Point 2 represents the elastic limit or proportional limit for the material being cut as well as for the tool material. The elastic limit is the point on the elastic curve (1–2) at which the force applied to the material is a maximum in terms of the fact that if the force is released, the material will return to its original state without permanent deformation. To be effective, the tool

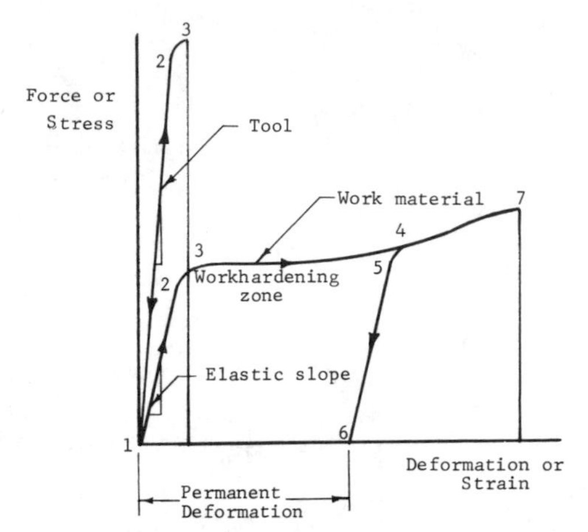

Fig. 2.12. Force–deformation or stress–strain diagrams comparing tool and workpiece materials.

material must obviously have a higher elastic limit than the material being cut.

Point 3 represents the yield point, the point at which the material significantly deforms as a result of a small change in applied force. Beyond point 3 is the plastic deformation range. Between points 3 and 4, small changes in force result in large deformations. During plastic deformation, the material work hardens, begins to lose its plastic capacity, stiffens to additional deformation, and finally ruptures at point 7. If the stress is released before rupture, as shown by point 4 in Fig. 2.12, the grain descends with decreasing stress along the elastic curve 5–6. At point 6 there is no stress on the grain, but it has been permanently deformed because of the plastic deformation that took place in the work hardening zone 3–4.

2.4 BASIC CONCEPTS OF STRESS AND STRAIN

A graph similar to that shown in Fig. 2.12 can be compiled by conducting a standard tensile test experiment. A specimen of the type shown in Fig. 2.13 is used in this type of test. By placing a load on the specimen and measuring the gauge length, common physical properties of materials such as tensile strength and modulus of elasticity can be determined.

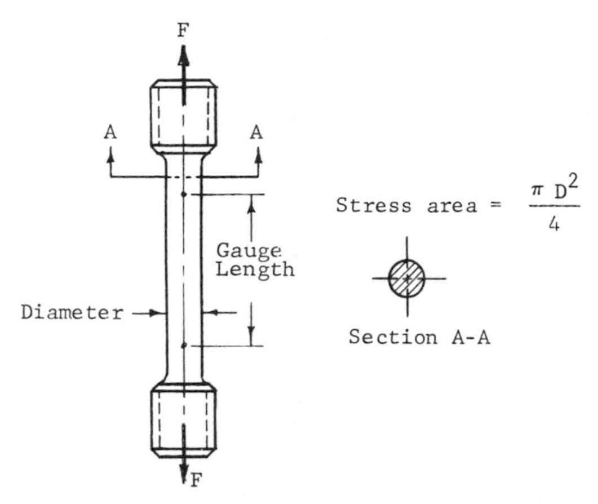

Fig. 2.13. Standard tensile test specimen.

If a force **F** is placed on the specimen, it will be absorbed and distributed internally. The internal mechanical reaction to the external force **F** can be detected in the form of stretching of the specimen. This deformation can then be recorded as a function of **F**. A measure of the internal distribution of the force is called "stress" and can be written as

$$S = \frac{\mathbf{F}}{A} \tag{2.7}$$

where S is the stress (lb/in.^{2}), **F** is the force (lb), and A is the area (in.^{2}).

When a material is subjected to a load, it deforms in a fashion that is dependent on the type of load. The test specimen of Fig. 2.13 exposed to an external tensile load will stretch. This will result in a change in the gauge length. A sensitive displacement instrument known as an *extensometer* can accurately measure this change. The change in length per unit length caused by a stress is defined as the *strain*. It can be written as

$$e = \frac{\Delta L}{L} \tag{2.8}$$

where e is the strain (in./in.), ΔL is the change in gauge length (in.), and L is the gauge length (in.).

A further examination of Fig. 2.12 reveals a region between points 1 and 2 where the relationship between the stress and strain is linear. It is described here by a straight line and is referred to as the *elastic region*. Within this elastic region, the ratio of the stress to the corresponding strain is constant. This constant ratio is defined as the *modulus of elasticity*. Analytically it can be expressed as

$$E = \frac{S}{e} \tag{2.9}$$

where E is the modulus of elasticity (lb/in.^{2}), S is the stress (lb/in.^{2}), and e is the strain (in./in.). By substituting terms from Eqs. 2.7 and 2.8 into Eq. 2.9, the modulus of elasticity can be written as

$$E = \frac{\mathbf{F}(L)}{A(\Delta L)} \tag{2.10}$$

EXAMPLE 2.8. Calculate the stress in the $\frac{1}{2}$-in.(12.7-mm)-diameter test specimen shown in Fig. 2.13 with an external force of 1000 lb (4448 N) acting on it.

From Eq. 2.7,

$$S = \frac{4\mathbf{F}}{\pi D^2}$$

where $A = \frac{\pi D^2}{4}$. Substituting yields

$$S = \frac{4(1000)}{3.14(0.5)^2}$$

$$= 5096 \text{ lb/in.}^2 \ (35.14 \times 10^6 \text{ Pa})$$

EXAMPLE 2.9. Calculate the strain in the specimen described in example 2.8 if a change of length of 0.00034 in. (0.0086 mm) was measured over a gauge length of 2 in. (50.8 mm) for the corresponding load of 1000 lb (4448 N).

Substituting into Eq. 2.8 yields

$$e = \frac{0.00034}{2.0}$$

$$= 1.7 \times 10^{-4} \text{ in./in.} \ (1.7 \times 10^{-4} \text{ m/m})$$

EXAMPLE 2.10. Taking into account the data from examples 2.8 and 2.9, solve for the modulus of elasticity of the test specimen.

Substituting numerical values in Eq. 2.10 gives

$$E = \frac{1000(2.0)}{(1.96 \times 10^{-1})(3.4 \times 10^{-4})}$$

$$= 30 \times 10^6 \text{ psi} \ (206.84 \times 10^9 \text{ Pa})$$

This is the modulus of elasticity for steel.

In Fig. 2.11, the plastic shear zone labeled 3–4 is a region where a grain has acting on it forces applied in a fashion that result in an angular deformation. This shearing action can be described by equations that are similar in form to those developed for the tensile model. A simplified cubic model of the deformation of a specimen as a result of a shearing force is given in Fig. 2.14.

If a shearing force couple ($F_s \times L_s$) is placed on the cubic model, it will be absorbed and distributed internally. The internal mechanical reaction to the external shearing force ($\mathbf{F}_s$) can be detected in the form of an angular deformation of the specimen. This angular deformation is indicated by the symbol e_s and is measured in terms of radians for small angular displacements below the proportional limit. From Fig. 2.14 it can be seen that

$$\frac{\Delta L_s}{L_s} = \tan e_s$$

where e_s is given in degrees. But, for small angles,

$$\tan e_s = e_s\,(\text{rad}) = \sin e_s$$

Therefore,

$$\frac{\Delta L_s}{L_s} = e_s\,(\text{rad})$$

For convenience, the shear strain can be defined as

$$e_s = \frac{\Delta L_s}{L_s} \tag{2.11}$$

where e_s is the shear strain (in./in. or rad), ΔL_s is the shear deformation (in.), and L_s is the shear length (in.). In a manner similar to that used in the tension model, the shear stress can be written as

$$S_s = \frac{\mathbf{F}_s}{A_s} \tag{2.12}$$

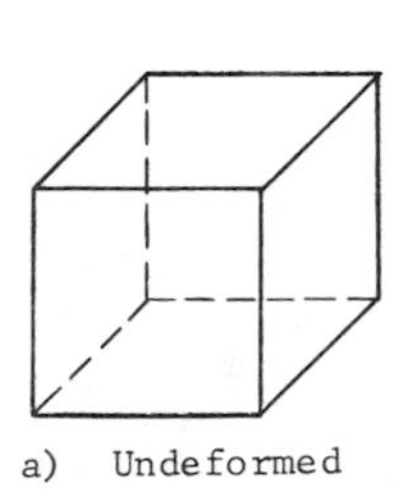

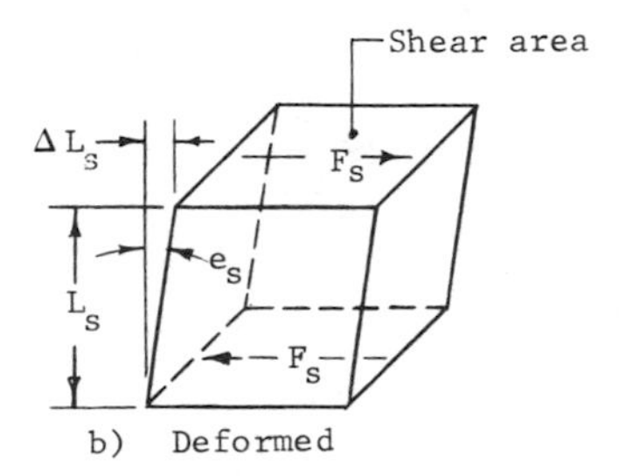

Fig. 2.14. Effect of shearing force on cubic model. (a) Undeformed model. (b) Deformed model.

where S_s is the shear stress (lb/in.^{2}), $\mathbf{F}_s$ is the shear force (lb), and A_s is the shear area (in.^{2}).

If a test were conducted where a shear force was applied to a cubic model of the type shown in Fig. 2.14 and the corresponding deflection were measured, a shear-stress–shear-strain diagram of the type shown in Fig. 2.12 could be compiled. A linear elastic region where the shear-stress–shear-strain ratio is constant would be of the type indicated between points 1 and 2. The slope of this straight line is called the *shear modulus of elasticity*. It can be expressed as

$$G = \frac{S_s}{e_s} \tag{2.13}$$

where G is the shear modulus of elasticity (lb/in.^{2}), S_s is the shear stress (lb/in.^{2}), and e_s is the shear strain (in./in.). By substituting terms from Eqs. 2.11 and 2.12 into Eq. 2.13, the shear modulus can be written as

$$G = \frac{\mathbf{F}_s(L_s)}{A(\Delta L_s)} \tag{2.14}$$

In examining Fig. 2.12, a comparison of physical properties for tool and for workpiece materials becomes evident. Table 2.1 lists a series of typical physical properties for some materials. As can be seen, tungsten, a tool material, has a much higher tensile strength than mild steel, a common workpiece material. The modulus of elasticity of tungsten is also higher than that of mild steel. In the event of tests being conducted on tungsten and mild steel for determination of physical properties, the expectation is that curves of the types displayed in Fig. 2.12 would result. Another point of interest is the comparison of the modulus of elasticity (tension and compression) with the shear

Table 2.1. Typical Physical Properties of Some Materials

Material	Density (lb/in.^{3})	Modulus of Elasticity, E (psi)	Shear Modulus of Elasticity, G (psi)	Tensile Yield Stress (psi)	Tensile Ultimate Stress (psi)
Aluminum	0.097	10×10^6	4×10^6	20×10^3	30×10^3
Brass	0.31	14×10^6	5.5×10^6	25×10^3	60×10^3
Iron, cast	0.28	15×10^6	6×10^6	6×10^3	20×10^3
Steel, mild	0.283	30×10^6	12×10^6	35×10^3	60×10^3
Steel, high-strength	0.283	30×10^6	12×10^6	120×10^3	180×10^3
Titanium	0.162	17×10^6	7×10^6	140×10^3	160×10^3
Tungsten	0.68	50×10^6	20×10^6	220×10^3	220×10^3

modulus of elasticity. The shear modulus of elasticity is lower than that for tension and compression. The shear yield stress is also usually lower than the tensile yield stress. As an example, experiments have shown that for a material such as structural steel, the yield point in shear is in the range of 0.55–0.60 of the yield point in tension.[1] As a result, it can be written that for steel,

$$S_{yps} = 0.55 S_{ypt} \tag{2.15}$$

where S_{yps} is the stress yield point in shear (lb/in.^{2}) and S_{ypt} is the stress yield point in tension (lb/in.^{2}).

EXAMPLE 2.11. Calculate the expected elongation of a 3-ft(0.914-m)-long, $\frac{3}{4}$-in. (19.05-mm)-diameter steel rod as a result of a 10,000-lb (44,482-N) load acting on it in an axial direction.

From Eq. 2.10,

$$\Delta L = \frac{\mathbf{F}(L)}{A(E)}$$

$$= \frac{10{,}000(36)}{0.44(30 \times 10^6)}$$

$$= 0.027 \text{ in. } (0.6858 \text{ mm})$$

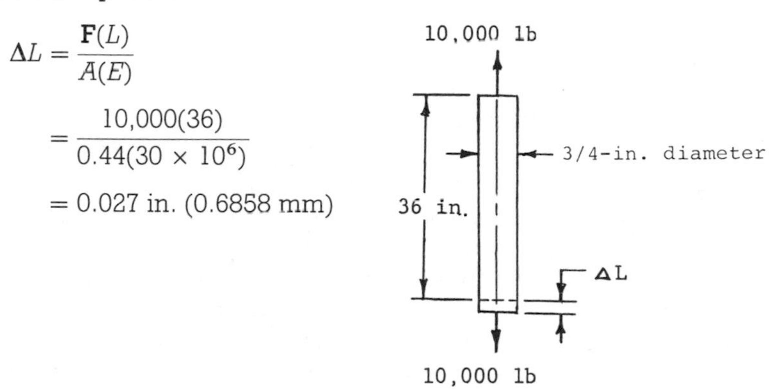

2.5 EXPERIMENTAL DETERMINATION OF PHYSICAL PROPERTIES

Physical properties of materials are determined by experimental tests. Figure 2.15 lists examples of how a tensile, a compressive, and a shear stress can be applied to a body. Of interest is the fact that if an experiment were conducted and the load (force) and deformation data were compiled, curves of the type shown in Fig. 2.12 could be plotted for the tensile, compressive, and shear stresses. From the graphed data, physical properties such as tensile strength, modulus of elasticity, and ductility could be determined. It should also be noted that when

[1] S. Timoshenko, *Strength of Materials*, Van Nostrand, New York, 1955, p. 62.

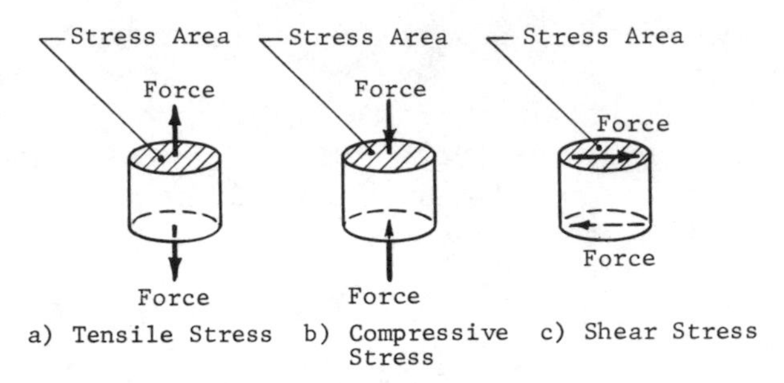

Fig. 2.15. Tensile (a), compressive (b), and shear (c) stresses.

comparing the tensile, compressive, and shear cases, similarities would exist in the general shape of the curves, although specific values would be different.

As an example of how some physical properties can be determined, let us assume that experiments were conducted on tensile, compressive, and shear specimens made from steel. For the sake of convenience, let us further assume that in all cases the stress area is 1 in.^{2} and that the gauge length is 1 in. long, as indicated in Table 2.2.

A compilation of the measurements of the deformations corresponding to applied loads that were placed on the test specimens is shown in Table 2.3. A plot of the data in terms of stress and strain is given in Fig. 2.16. An inspection of the diagram of the experimental data can lead to an evaluation of certain physical properties. As an example, the modulus of elasticity can be calculated by measuring the slope of the curves in the linear elastic region. It can then be seen that for tension, the modulus of elasticity can be evaluated from Eq. 2.10 as

$$E_t = \frac{\mathbf{F}}{A} \times \frac{L}{\Delta L} = \frac{30{,}000}{0.001} = 30 \times 10^6 \text{ psi } (206 \times 10^9 \text{ Pa})$$

In a similar fashion, for compression

$$E_c = \frac{\mathbf{F}}{A} \times \frac{L}{\Delta L} = \frac{30{,}000}{0.001} = 30 \times 10^6 \text{ psi } (206 \times 10^9 \text{ Pa})$$

Table 2.2. Convenient Test Specimen Data

Test	Stress Area (in.^{2})	Gauge Length (in.)
Tension	1	1
Compression	1	1
Shear	1	1

Table 2.3. Load–Deformation Data on Steel

Load (lb)	Deformation Tension (in.)	Compression (in.)	Shear (in.)
0	0.00000	0.00000	0.00000
10,000	0.00033	0.00033	0.00083
20,000	0.00067	0.00067	0.00167
22,000	0.00073	0.00073	0.00200
24,000	0.00080	0.00080	0.00300
25,000	0.00083	0.00083	0.00400
25,000	0.00083	0.00083	0.00500
30,000	0.00100	0.00100	
35,000	0.00116	0.00116	
37,500	0.00135	0.00135	
40,000	0.00200	0.00200	
43,000	0.00300	0.00300	
45,000	0.00400	0.00400	
43,000	0.00500	0.00500	

Also, for shear

$$G = \frac{\mathbf{F}}{A} \times \frac{L}{\Delta L} = \frac{20{,}000}{0.00167} = 11.98 \times 10^6 \text{ psi } (82.6 \times 10^9 \text{ Pa})$$

It is noted that A and L were set at values of 1 for convenience in calculating.

The stress at the highest point of the elastic region (top of straight line) is called the *proportional limit stress*. From Fig. 2.16 it can be seen that for tension the proportional limit stress is equal to

$$S_{\text{plt}} = \frac{\mathbf{F}_t}{A_t} = 35{,}000 \text{ psi } (241 \times 10^6 \text{ Pa})$$

Further examination of Fig. 2.16 reveals that for compression the proportional limit stress is

$$S_{\text{plc}} = \frac{\mathbf{F}_c}{A_c} = 35{,}000 \text{ psi } (241 \times 10^6 \text{ Pa})$$

and for shear, the proportional limit stress is

$$S_{\text{pls}} = \frac{\mathbf{F}_s}{A_s} = 20{,}000 \text{ psi } (137.9 \times 10^6 \text{ Pa})$$

Beyond the proportional limit stress, it is observed that the strain increases considerably without corresponding increases in stress. The

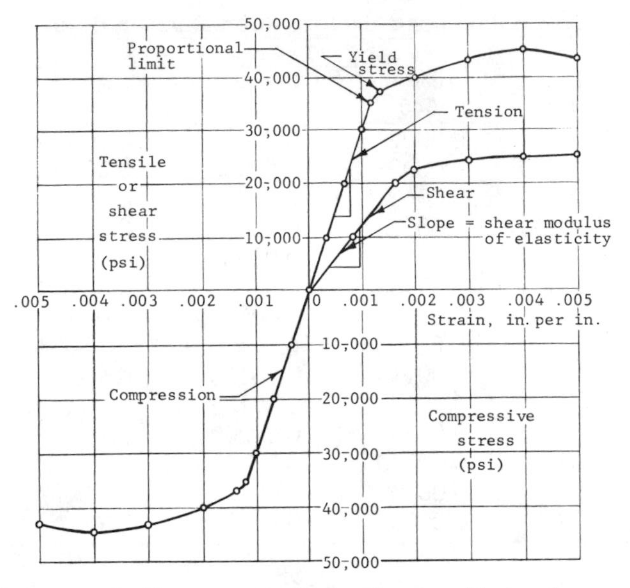

Fig. 2.16. Stress-strain diagrams for steel with regard to tension, compression, and shear.

stress at which this action commences is called the *yield stress*. From Fig. 2.16, these yield point stresses can be determined. They are equal to

$$S_{\mathrm{ypt}} = \frac{\mathbf{F}_t}{A_t} = 37{,}000 \text{ psi } (255 \times 10^6 \text{ Pa}) \text{ for tension}$$

$$S_{\mathrm{ypc}} = \frac{\mathbf{F}_c}{A_c} = 37{,}500 \text{ psi } (259 \times 10^6 \text{ Pa}) \text{ for compression}$$

$$S_{\mathrm{yps}} = \frac{\mathbf{F}_s}{A_s} = 22{,}000 \text{ psi } (152 \times 10^6 \text{ Pa}) \text{ for shear}$$

Ductility and brittleness are properties of tools and workpiece materials that are affiliated with hardness. Ductility gives a measure of the ability of a material to stretch or deform when it is stressed beyond the elastic limit. On the other hand, the opposite of ductility is brittleness, which usually indicates that a material will rupture (fracture) with little permanent distortion. Going back to Fig. 2.12, the tool curve indicates a brittle material, whereas the work material curve indicates a ductile material. A quantitative measure of the degree of

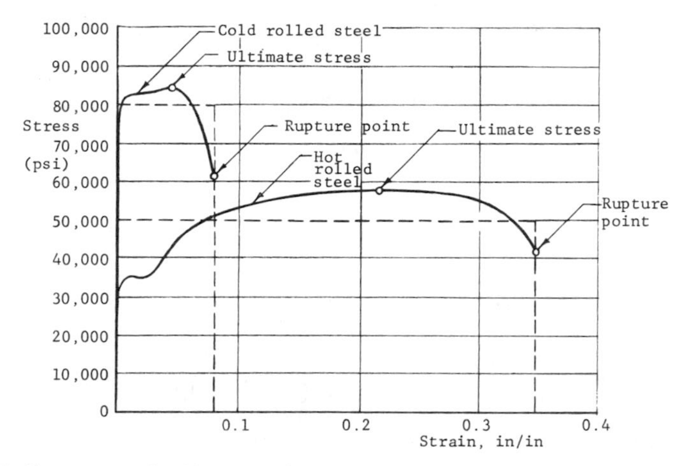

Fig. 2.17. Stress–strain diagrams for cold-rolled and hot-rolled steel compiled from a tensile specimen.

ductility can be obtained by measuring the area under the stress–strain diagram.

An interesting comparison of ductility with regard to a cold-rolled steel and a hot-rolled steel is given in the stress–strain diagrams shown in Fig. 2.17.[2] The area represented by the dashed lines approximates the area under the stress–strain curve. This area is a measure of the amount of work (force × displacement) or energy that a material can absorb before rupture. This area also gives a quantitative measure of the brittleness or ductility of the material. It can be seen that the yield point for the cold-rolled steel is substantially higher than that for the hot-rolled steel. However, an examination of the area under the individual curves shows that the elongation (stretch) of the cold-rolled steel is substantially lower. A measure of the relative brittleness or relative ductility can be made by evaluating the area under these stress–strain diagrams. This indicates a direct value of the amount of energy or work (in.-lb) that the test specimen absorbed per unit volume before it ruptured (failed).

As indicated in Fig. 2.17, a rectangle of dimensions 80,000 psi by 0.08 in./in. approximates the area under the cold-rolled-steel stress–strain diagram. For convenience of the calculation, assume that the test specimen is a cube, that is, the stress area is 1 in.^{2} and the gauge

[2] Alden W. Counsell, Test for True Stress and Strain, unpublished laboratory experiment, Northeastern University, Boston, 1949.

length is 1 in. The work (force × displacement) or energy expended on the specimen can be written as

$$\text{Work} = (\text{Stress} \times \text{stress area})\,(\text{Strain} \times \text{length of specimen})$$
$$= \text{Force} \times \text{displacement}$$

Thus,

$$W = \mathbf{F} \cdot \mathbf{D} \tag{2.16}$$

Substituting values from Fig. 2.17 yields

$$W = 80{,}000 \text{ lb} \times 0.08 \text{ in.}$$
$$= 6400 \text{ in.-lb } (722 \text{ N-m})$$

Similarly, the area under the hot-rolled-steel stress–strain diagram can be approximated by a rectangle of dimensions 50,000 psi by 0.345 in./in. Using Eq. 2.16, the work or energy absorbed by the specimen to fracture can be evaluated by writing

$$W = \mathbf{F} \cdot \mathbf{D}$$
$$= 5000 \text{ lb} \times 0.345 \text{ in.}$$
$$= 17{,}250 \text{ in.-lb } (1946 \text{ N-m})$$

Note that units of pounds and inches are used insofar as these are interchangeable with units of psi and in./in. for the case where the stress area is 1 in.^{2} and the gauge length is 1 in. long. If the stress area were to be other than 1 in.^{2} and the gauge length other than 1 in. long, as is the case with a standard test specimen as shown in Fig. 2.13, the work can be expressed in terms of stress and strain as

$$\frac{\text{Work}}{\text{volume}} = \text{Stress} \times \text{strain}$$

or

$$\frac{W}{V} = S \times e \tag{2.17}$$

Again, examining the areas in Fig.. 2.17 under the stress–strain diagrams, which are represented by a rectangular shape, yields

$$\frac{W}{V} = 80{,}000 \text{ lb/in.}^2 \times 0.08 \text{ in./in.}$$
$$= 6400 \text{ in.-lb/in.}^3 \ (44 \times 10^6 \text{ N-m/m}^3) \text{ for cold-rolled steel}$$

and

$$\frac{W}{V} = 50{,}000 \text{ lb/in.}^2 \times 0.345 \text{ in./in.}$$

$$= 17{,}250 \text{ in.-lb/in.}^3 (119 \times 10^6 \text{ N-m/m}^3) \text{ for hot-rolled steel}$$

As can be seen from the data compiled, it takes substantially more work (energy) to fracture hot-rolled steel than cold-rolled steel. This is the case despite the fact that the cold-rolled steel has a much higher tensile strength. However, its deformation prior to fracture is substantially less, yielding a smaller area under the stress–strain curve. A logical conclusion to reach in comparing Fig. 2.17 with Fig. 2.12 is that there is a similarity. Since the cold-rolled steel has the higher tensile strength, it should be able to stress the hot-rolled steel beyond its yield point and cause it to flow plastically. As a result, one may be inclined to state that the cold-rolled steel could be used as a tool to machine the hot-rolled steel. This is true if the physical properties do not change. However, under the harsh conditions of metal cutting, not only is heat generated, which affects the physical properties, but also the machined part undergoes work hardening due to plastic flow, which also affects the physical properties. The net result is that the cold-rolled tool does not perform effectively.

EXAMPLE 2.12 Calculate the work per unit volume required to fracture a tungstenlike material with a modulus of elasticity of 50×10^6 psi (345×10^9 Pa). It is noted that the specimen fractures at a proportional elastic limit stress of 210×10^3 lb/in.^{2} (1.47×10^9 Pa).

The resultant strain is determined from Eq. 2.9 as

$$e = \frac{S}{E} = \frac{210 \times 10^3}{50 \times 10^6}$$

$$= 4.2 \times 10^{-3} \text{ in./in. } (4.2 \times 10^{-3} \text{ m/m})$$

It is noted that the area under the stress–strain diagram within the elastic region is triangular in shape. It can be written in the form of Eq. 2.17 as

$$\text{Area}_{s-s} = \frac{W}{V} = \tfrac{1}{2}(S \times e)$$

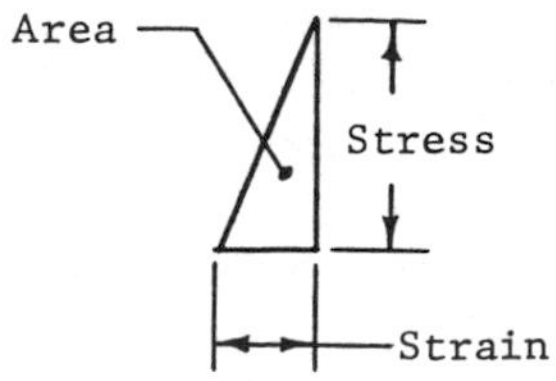

Area
Stress
Strain

The coefficient $\frac{1}{2}$ accounts for the triangular shape. Equation 2.17 is based on the assumption that the area under the stress–strain diagram is approximated by a rectangular shape. Solving for the work done yields

$$\frac{W}{V} = \tfrac{1}{2}(210 \times 10^3) \times (4.2 \times 10^{-3})$$

$$= 442 \text{ in.-lb/in.}^3 \, (3.05 \times 10^6 \text{ N-m/m}^3)$$

2.6 MECHANICAL EQUIVALENT OF HEAT

Metal cutting produces heat as a result of the expenditure of energy in order to perform the operation. This is most evident in the high temperature of the chip as it is being cut from the workpiece. The temperature of the chip is elevated by hundreds of degrees Fahrenheit as it is being formed, as can be attested to by anyone who has been burnt by one. The temperature of the chip can, on occasion, be estimated by visual observation. This can be especially done on steel by noting the color change of the chip. Blue chips, for example, have been elevated to the 600°F range as indicated by so-called temper colors. On steel, these colors are due to oxidation at elevated temperatures that result in the formation of colored films of varying thickness of iron oxide on the surface of the chip.

Another case of interest dealing with the temperature of chips is the observation of hardened steel being machined with a carbide tool. This process can produce a cherry-red hot chip, which by color matching can be estimated to be elevated to the 1400°F temperature range. The generation of heat is very apparent in the metal cutting process, and this observation led to experimentation relating mechanical work and heat.

Of special interest is that early experiments dealing with mechanical work and heat were conducted on the metal cutting operation. Some of these early experiments were performed by Count Rumford (1753–1814), an American-born scientist. His name was Benjamin Thompson, and he lived in Europe after the Revolutionary War. He gained the title of count in 1791 for his work in improving the living conditions of the poor in Munich.

While engaged as superintendent over the boring of cannons at a military arsenal in Munich, he was impressed by the rapid increase in the temperature of a brass cannon that was being bored. As a result, he conducted an experiment. He placed a blunt borer forced against

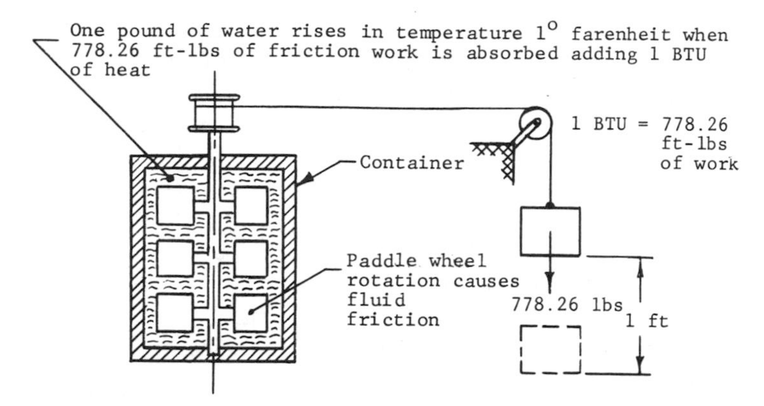

Fig. 2.18. Symbolic representation of an experiment showing the mechanical equivalent of heat.

the bottom of a cylinder in a container filled with water. He then began to rotate the cylinder by a force provided by horses. To measure the temperature of the water, he used a mercury thermometer. The data he collected were that of the temperature of the water as a function of time. Starting with 60°F water, he found that in 1 hr, during which time the cylinder was rotated against the borer, the temperature of the water rose to 107°F. In $2\frac{1}{2}$ hr, he was able to boil the water.[3] Many observers of the experiment at the time were astonished that a large quantity of water could be brought to the boiling point without a fire.

Count Rumford's early work ultimately led to experimentation involving the precise measurement of the mechanical equivalent of heat. James P. Joule (1818–1889), an English physicist, conducted the famous experiment dealing with the measurement of the mechanical equivalent of heat. He placed water in a container and agitated it, causing fluid friction that raised the temperature of the water. Figure 2.18 symbolizes the results of the experiment, with an updated value of the mechanical equivalent of heat.

The conclusion that Joule reached was that the quantity of heat required to elevate the temperature of 1 lb of water 1°F called upon the expenditure of 781.5 ft-lb of mechanical energy. Today, the accepted standard is 778.26 ft-lb, as shown in Fig. 2.18. This amount of energy is defined as one British thermal unit (Btu). It is the amount of energy required to raise the temperature of 1 lb of water from 59° to 60°F at 1 atm pressure. At a result of his famous experiment, Joule is credited

[3] Charles G. Frazer, *Half-Hours with Great Scientists*, Reinhold, New York, 1948, p. 343.

with the discovery of the first law of thermodynamics. This law states that whenever energy is transformed from one form to another, it is always conserved. In other words, energy can neither be created nor destroyed. Analytical statements of the first law are written in the form that heat and work are mutually convertible and that the total energy associated with energy conservation remains constant.

The availability of the mechanical equivalent of heat enables calculations to be performed on the temperature rise that takes place in a chip during the metal cutting operation. In preparation for this calculation, let us review some basic engineering terms. The work performed by a force moving through a displacement in the same direction as the force is defined as

$$\text{Work} = \text{Force} \times \text{Displacement}$$

or

$$W = \mathbf{F} \cdot \mathbf{D}$$

It is noted that force and displacement are vector quantities. They are defined as quantities possessing both magnitude and direction. In order to add vectors, the parallelogram law must be applied. The direction as well as the magnitude of the vector must be taken into consideration. Figure 2.19 illustrates the addition of two vectors.

EXAMPLE 2.13. Given the two vectors shown in Fig. 2.19, solve for the resultant. Note $\mathbf{F}_1 = 20\angle 60°$ and $\mathbf{F}_2 = 40\angle 0°$.
Resolving the vectors into x and y components yields

$$F_{1x} = |\mathbf{F}_1| \cos\theta_1 = 20(0.5) = 10$$

$$F_{1y} = |\mathbf{F}_1| \sin\theta_1 = 20(0.866) = 17.32$$

$$F_{2x} = 40$$

$$F_{2y} = 0$$

Adding x and y components yields

$$R_x = F_{1x} +\!\!\!\rightarrow F_{2x} = 10 + 40 = 50$$

$$R_y = F_{1y} +\!\!\!\rightarrow F_{2y} = 17.32 + 0 = 17.32$$

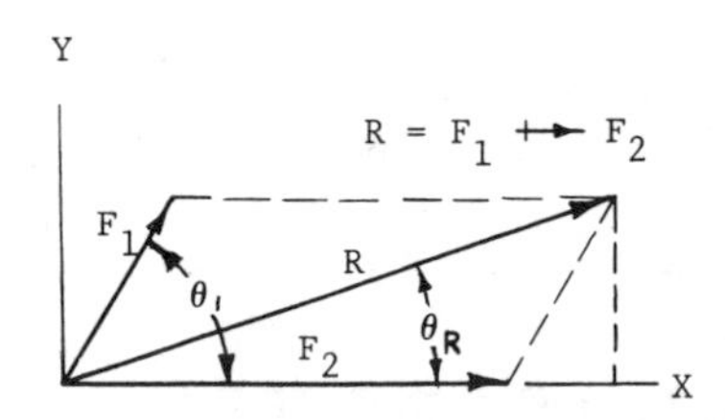

Fig. 2.19. Addition of two vectors.

and

$$|\mathbf{R}| = \sqrt{R_x^2 + R_y^2}$$

$$|\mathbf{R}| = 52.9 \text{ (magnitude)}$$

Direction of the resultant can be written as θ_R and

$$\tan\theta_R = \frac{R_x}{R_y} = \frac{50}{17.32} = 2.887$$

$$\theta_R = 70.89^\circ$$

For the sake of emphasizing the importance of taking into account the direction of a vector, Fig. 2.20 is given. It illustrates a force and an affiliated displacement, which is at an angle to the force. For convenience, the force vector **F** in Fig. 2.20 can be resolved into two components. F_x is the component in the x direction and F_y is the component in the y direction. The vector sum of these two components equals the force vector **F**, which is the resultant of F_x and F_y according to the parallelogram law for the addition of vectors.

The work done by the force **F** in Fig. 2.20 is the product of the force in the direction of the displacement and the displacement. The force and displacement must be in the same direction in order to calculate the work done by the force. The work can be written as

$$W = F_x \times \mathbf{D}_1$$

where

$$F_x = \mathbf{F}\cos\theta$$

and

$$W = \mathbf{F} \cdot \mathbf{D}_1 \cos\theta \tag{2.18}$$

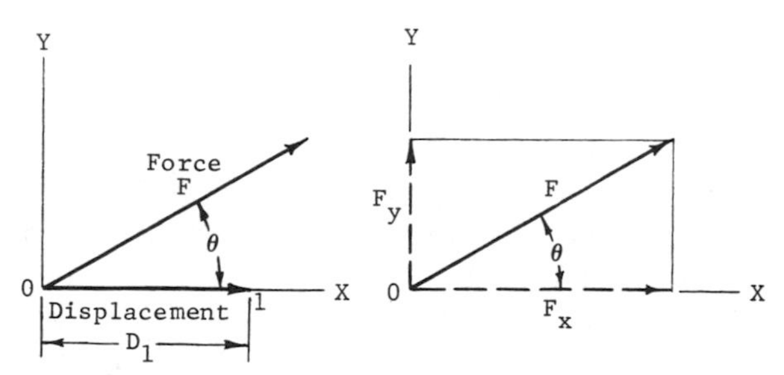

Fig. 2.20. Illustration of force–displacement relationship. (a) Force–displacement vectors. (b) Components of force.

Of special interest in Fig. 2.20 is that the force in the y direction (F_y) does no work because there is no affiliated displacement in the y direction.

EXAMPLE 2.14 A 100-lb (444.8-N) force directed at a 30° angle is required to drag a box a distance of 3 ft (0.914 m) as shown in the diagram below. Calculate the amount of work performed in moving the box. Also, determine how much heat is generated as a result of the expenditure of energy in moving the box.

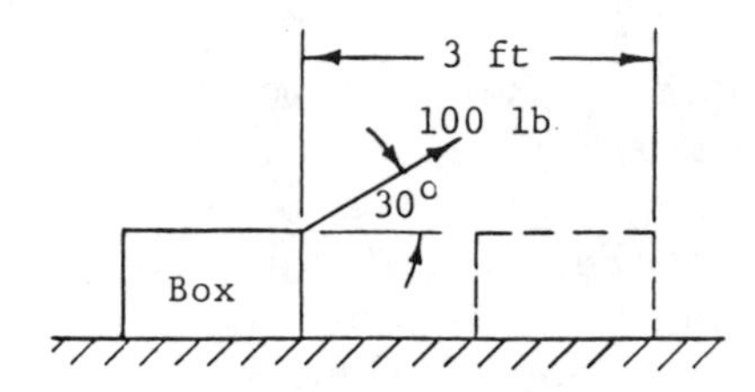

Substituting into Eq. 2.18, we obtain

$$\begin{aligned} W &= \mathbf{F} \cdot \mathbf{D}_1 \cos\theta \\ &= 100(3)(0.866) \\ &= 259.8 \text{ ft-lb } (352 \text{ N-m}) \end{aligned}$$

Using the mechanical equivalent of heat where 1 Btu = 778.26 ft-lb, we get

$$\begin{aligned} Q &= W \times \frac{1 \text{ Btu}}{778.26 \text{ ft-lb}} \\ &= \frac{259.8 \times 1}{778.26} \\ &= 0.334 \text{ Btu} \end{aligned}$$

When a chip is formed in the metal cutting operation, a force is applied that permanently deforms the work material, resulting in work being performed. This work is done in the plastic region of the stress–strain diagram, and it is not recoverable. It is lost because it is converted to heat that eventually is dissipated. Figure 2.21 shows examples of work being performed during (a) tension, (b) compression, and (c) shear. For purposes of convenience, the illustration is simplified by the assumption that the material deforms uniformly once it is stressed beyond the yield point. A further assumption is made that the material deforms uniformly under the yield stress despite the change that takes place in the cross section that represents the stress area. These assumptions simplify the model to one where the work can be calculated in terms of the rectangular area under the yield stress–strain diagram, an analysis similar to that represented by the dashed lines in Fig. 2.17.

As a preparation for the calculation of the temperature rise in the chip, let us now look at the simple cases shown in Fig. 2.21. Once again, for convenience of the calculation, another assumption is made. It is assumed that the specimens in Fig. 2.21 are cubes of 1-in. dimensions that are made from steel. For the tensile and compressive cases, the area under the stress–strain diagram is approximated to have an average yield stress of 45,000 psi (310×10^6 Pa); this places the specimens in the plastic state, during which they deform 0.3 in. (7.62 mm).

Solving for the work performed on the specimens yields

$$\text{Work/volume} = \text{Stress} \times \text{Strain}$$

For a cube where stress area = 1 in.^{2} (645 mm^2), gauge length = 1 in. (25.4 mm), and volume = 1 in.^{3} (16.4×10^3 mm^3), Eq. 2.17 reduces to Eq. 2.16 and can be written as

$$\text{Work} = \text{Force} \times \text{Displacement} = \mathbf{F} \cdot \mathbf{D}$$

or

$$\begin{aligned} W &= 45{,}000 \text{ lb} \times 0.3 \text{ in.} \\ &= 13{,}500 \text{ in-lb} = 1125 \text{ ft-lb } (1525 \text{ N-m}) \end{aligned}$$

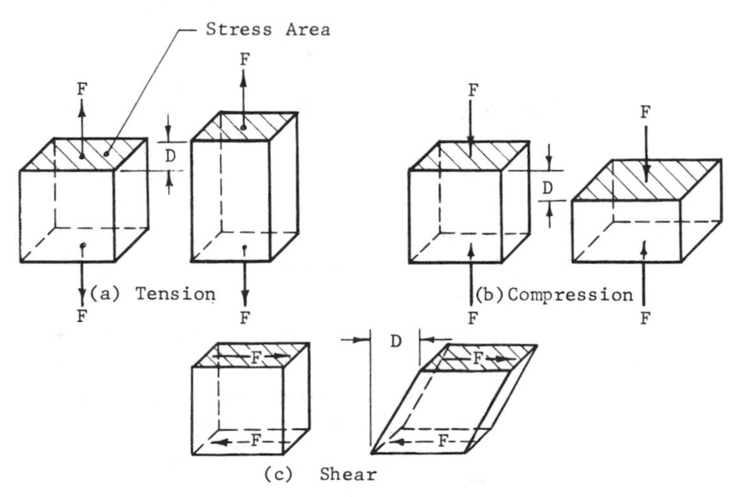

Fig. 2.21. Examples of work being done in (a) tension, (b) compression, and (c) shear.

By using the mechanical equivalent of heat, the work can be converted to thermal energy. As previously stated, a unit of measure of thermal energy is the British thermal unit or Btu. It is defined as the energy required to raise the temperature of 1 lb of water from 59° to 60°F at 1 atm pressure. In terms of mechanical units, one Btu is equal to 778.26 ft-lb of work. Converting the work to Btu's yields

$$W = 1125 \text{ ft-lb} \times \frac{\text{Btu}}{778.26 \text{ ft-lb}}$$

$$= Q = 1.4455 \text{ Btu}$$

where work (W) is written in terms of heat (Q).

Of special interest is the determination of the temperature rise of the specimen as a result of the 1125 ft-lb of work performed on it. If all of the work is converted to heat and all of the heat is retained in the specimen, the temperature rise can be written as

$$\Delta T = \frac{Q}{CW_t} \tag{2.19}$$

where ΔT is the change in temperature; Q is the heat; C is the specific heat or heat capacity (the amount of heat required to raise the temperature of a unit mass 1°F)—for steel, the value is 0.11 Btu/lb-°F; and W_t is the weight of material—for steel, it is 0.282 lb/in.^{3}. Substituting numerical values of this example in Eq. 2.19 yields

$$\Delta T = 1.4455 \text{ Btu} \times \frac{\text{lb-°F}}{0.11 \text{ Btu}} \times \frac{1}{0.282 \text{ lb}}$$

$$= 46.6\text{°F}$$

The temperature of the specimen rises 46.6°F when the work energy is converted to heat, and all of the heat is retained in the specimen. In the event that the specimen was originally at a room temperature of 70°F, an increase of 46.6°F would have elevated the temperature of the specimen to 116.6°F. Both the tensile and compressive specimens would experience this same temperature rise in converting the 1125 ft-lb (1525 N-m) of work to heat.

The shear specimen would also rise in temperature by 46.6°F if it had converted 1125 ft-lb (1525 N-m) of work to heat. Of interest is the evaluation of the shear stress and corresponding deformation that results in this same amount of work being expended. The relationship between stress yield points of tensile and shear steel specimens can be expressed in terms of Eq. 2.15 as

$$S_{\text{yps}} = 0.55 S_{\text{ypt}}$$

The comparable value of average yield shear stress that would produce a plastic flow would then be equal to

$$S_{yps} = 0.55 \times 45{,}000 = 24{,}750 \text{ psi } (170.6 \times 10^6 \text{ Pa})$$

In order to generate the same work as in the tensile and compressive cases, the corresponding shear deformation with the 24,750 psi (170.6×10^6 Pa) average shear yield stress acting on the 1-in. shear area would be equal to

$$\mathbf{D}_s = \frac{\text{Work}}{\text{Force}}$$

$$= \frac{13{,}500 \text{ in.-lb}}{24{,}750 \text{ lb}}$$

$$= 0.545 \text{ in. } (13.843 \text{ mm})$$

EXAMPLE 2.15. In a heading operation, a $\frac{3}{8}$-in.(9.5-mm)-diameter by $\frac{3}{4}$-in.(19.1-mm)-long portion of a steel rod is formed into a $\frac{3}{4}$-in.(19.1-mm)-diameter fastener head. The average force required to form the head is 18,000 lb (80,068 N). If all the work is converted to heat and retained in the head of the fastener, calculate the rise in temperature that takes place.

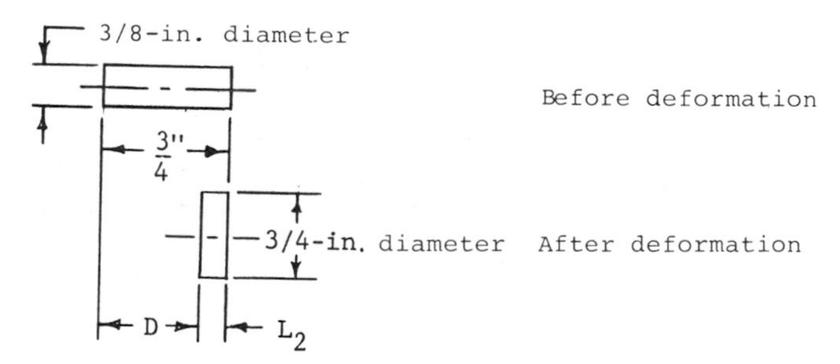

The distance over which the force acts is the difference in the length of the part.

$$\mathbf{D} = 0.75 - L_2$$

Since the volume of the head remains constant, then

$$\text{Vol } 1 = \text{Vol } 2$$

$$\left(\frac{\pi d_1^2}{4}\right)\frac{3}{4} = \left(\frac{\pi d_2^2}{4}\right)L_2$$

$$0.082 \text{ in.}^3 = 0.4416(L_2) \text{ in.}^3$$

$$L_2 = 0.1875 \text{ in. } (4.76 \text{ mm})$$

and

$$\mathbf{D} = 0.5625 \text{ in. } (14.3 \text{ mm})$$

Solving for the work performed, we obtain

$$W = \mathbf{F} \cdot \mathbf{D} = 18{,}000 \times 0.5625$$
$$= 10{,}125 \text{ in-lb} = 843.75 \text{ ft-lb } (1144 \text{ N-m})$$

Converting work to heat gives

$$Q = \frac{843.75}{778.26} = 1.084 \text{ Btu}$$

The weight of the head is

$$W_{\text{th}} = \frac{W_{\text{t}}}{\text{vol}} \times \text{vol} = 0.282 \times 0.0828$$
$$= 0.0233 \text{ lb } (0.0106 \text{ kg})$$

Solving for the rise in temperature by substituting values into Eq. 2.18 yields

$$\Delta T = 1.084 \text{ Btu} \times \frac{\text{lb-°F}}{0.11 \text{ Btu}} \times \frac{1}{0.0233 \text{ lb}}$$
$$= 423\text{°F}$$

EXAMPLE 2.16. A 1-in. (25.4 mm) cube is deformed as shown below by an average 24,750-lb (110,093-N) force. Calculate the temperature rise in the cube if all the work is converted to heat and retained by the cube.

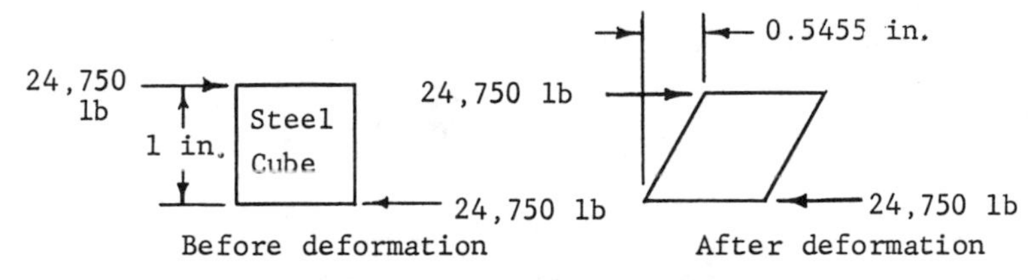

The work performed on the specimen is equal to

$$W = \mathbf{F} \cdot \mathbf{D} = 24{,}570 \times 0.5455$$
$$= 13{,}501 \text{ in.-lb} = 1125 \text{ ft-lb } (1525 \text{ N-m})$$

Converting work to heat yields

$$W = Q = \frac{1125}{778.26} = 1.4455 \text{ Btu}$$

Finally, the rise in temperature can be calculated by substituting numerical values into Eq. 2.18:

$$\Delta T = \frac{1.4455}{0.11 \times 0.282} = 46.6°\text{F}$$

2.7 SIMPLIFIED MODEL OF THE CHIP-FORMING OPERATION

An examination of the chip-forming operation indicates the occurrence of a large deformation in the original work material. A microscopic observation reveals that the deformation is in the form of a shear plastic flow that is confined to a well-bounded region between the workpiece and the chip. The chip is formed through this shear plastic deformation that causes the chip to emerge substantially thicker than the feed machine setting from which it was cut. The severity of the high stress induced to cause the chip to flow is reflected in terms of the high temperature of the formed chip. This bears witness to the amount of energy that must be expended to form the chip. Flow lines on the chip further document that the plastic deformation is at an angular orientation to the sides of the chip.

Figure 2.22 illustrates a simplified model of the chip-forming operaeration. This convenient model lends itself readily to a simple analysis of the work performed on a segment of the workpiece, which undergoes the permanent deformation leading to the chip formation. Segment 1 depicts the unstrained portion of the workpiece as it approaches the shear zone, where permanent deformation takes place. As the segment enters the narrow shear zone designated by the shear angle, a wiping plastic shear deformation begins to transform the segment into the chip. Segment 2 details the deformation that takes place along the shear angle (α), causing a progressive change to take place in the original square segment. As the segment continues to be absorbed into the shear zone, it deforms as indicated by segment 3. Continued deformation is illustrated by segment 4. Finally, the original square segment completely emerges from the shear zone and the deformation is concluded. The resultant shape is that of the form indicated by segment 5, an elongated parallelogram.

A comparison of segment 1 with segment 5 shows a large deformation that could be achieved by controlling forces acting in a variety of fashions. As an example, tensile forces could stretch segment 1 into segment 5 if these forces were so directed as to keep the segment

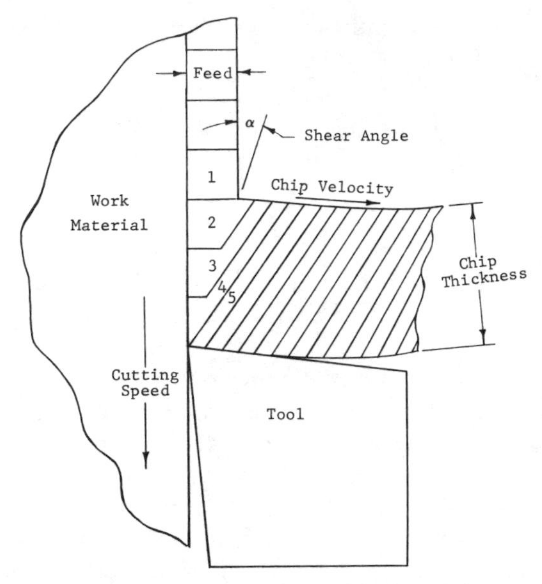

Fig. 2.22. Simplified model of the chip-forming operation.

in the plastic state during the deformation. The same could be said for the compression case. If compressive forces were so directed as to force a plastic flow, segment 1 could be squeezed into the shape of segment 5. In both cases, if the average force causing the plastic flow was multiplied by the deformation, it would provide a measure of the amount of work required to perform the deformation. This, in turn, could be converted to the heat equivalent, and the temperature of the segment could be calculated.

An example of a calculation of this type is now presented for the case where the assumption is made that all of the deformation is due to shear plastic flow. For purposes of analysis, an assumed model of the work material, shear zone element, and chip is given in Fig. 2.23. The stationary shear zone element is shown with the shear forces acting on it, which result in a displacement **D**. To enable a numerical analysis to take place, a further assumption is made that the model represents the flow lines in the metal. Corresponding numerical values are given as

Feed $= f = 0.021$ in. (0.533 mm)

Depth of cut $= d = 0.100$ in. (2.54 mm)

Width of shear zone in x direction $= a = \dfrac{f}{3} = 0.007$ in. (0.178 mm)

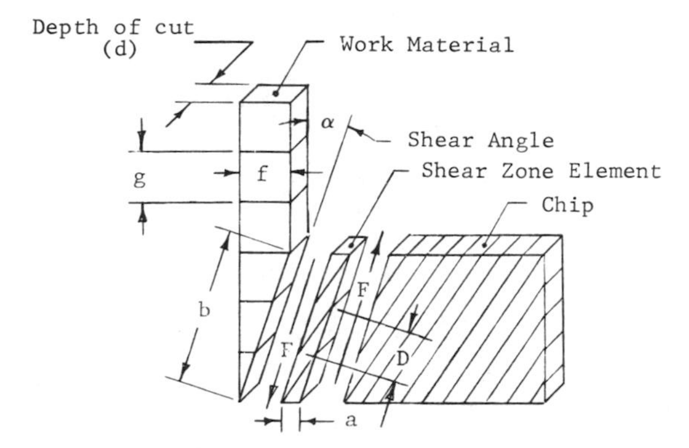

Fig. 2.23. Shear stress acting on shear zone element.

Yield shear stress = 25,650 psi (177 × 10^6 Pa)

Work material = Steel

The shear angle can be calculated from the geometry of the model. It can be written that

$$\tan \alpha = \frac{f}{3g} = \tfrac{1}{3}$$

where $f = g$ and $\alpha = 18.43°$. The work done on the shear zone element can be evaluated by measuring the permanent deformation that took place in the shear zone and multiplying this deformation by the average shear force acting on the shear area.

Let us now isolate that portion of the segment that is in the shear zone and determine its dimensions. This reduces the problem to one similar to that represented in Fig. 2.21(c), as can be seen in Fig. 2.24. The length of the shear section (b) can be written as

$$b = \frac{3g}{\cos \alpha}$$

$$= \frac{0.063}{0.9487}$$

$$= 0.0664 \text{ in. } (1.687 \text{ mm})$$

Inspection of the shear element shown in Fig. 2.24 reveals that the shear displacement **D** is equal to

$$\mathbf{D} = \frac{b}{3}$$

$$= 0.0221 \text{ in. } (0.561 \text{ mm})$$

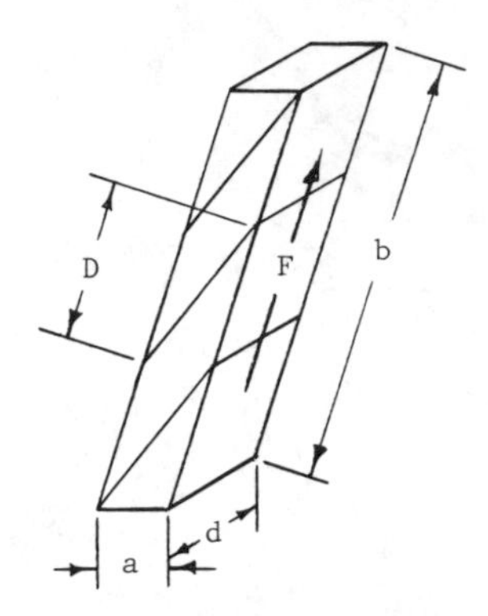

Fig. 2.24. Diagram of shear element.

The shear stress area A_{st} over which the shear force **F** acts can now be evaluated as

$$\begin{aligned} A_{st} &= bd \\ &= 0.0664 \times 0.100 \\ &= 0.00664 \text{ in.}^2 \ (4.28 \text{ mm}^2) \end{aligned}$$

The shear force can be determined under the assumption that all of the deformation takes place under the given yield shear stress. As a result,

$$\begin{aligned} \mathbf{F} &= \text{Stress} \times \text{Area} \\ &= 25{,}600 \times 0.00664 \\ &= 170 \text{ lb } (756 \text{ N}) \end{aligned}$$

With the shear force and shear displacement evaluated, the work performed on the shear element represented in Fig. 2.24 can now be determined as

$$\begin{aligned} W &= \mathbf{F} \cdot \mathbf{D} \\ &= 170 \times 0.0221 \\ &= 3.757 \text{ in.-lb} = 0.313 \text{ ft-lb } (0.424 \text{ N-m}) \end{aligned}$$

Converting the work to heat gives the generation of

$$Q = \frac{0.313}{778.26} = 4.023 \times 10^{-4} \text{ Btu}$$

The volume of the shear zone element that absorbs the heat generated by the work is equal to

$$\begin{aligned} \text{Vol} &= (a\cos\alpha) \times b \times d \\ &= 0.0066 \times 0.0664 \times 0.100 \\ &= 4.3824 \times 10^{-5}\ \text{in.}^3\ (0.718\ \text{mm}^3) \end{aligned}$$

Finally, solving for the temperature rise due to the work done on the shear element, we obtain

$$\Delta T = \frac{Q}{C \times W_t}$$

where

$$W_t = \text{Vol} \times \text{weight density}$$

$$W_t = 4.3824 \times 10^{-5} \times 0.282$$

$$W_t = 1.236 \times 10^{-5}\ \text{lb}$$

$$C = 0.11\ \text{Btu/lb-°F (for steel)}$$

After substituting, we get

$$\Delta T = \frac{4.023 \times 10^{-4}\ \text{Btu-lb-°F}}{0.11\ \text{Btu} \times 1.236 \times 10^{-5}\ \text{lb}}$$

$$= 295.9\text{°F}$$

In the event that all of the heat is retained in the chip, which was originally at a temperature of 70°F, the chip temperature would rise to 365.9°F as a result of the work performed in the shear zone by the shear forces.

2.8 DERIVATION OF SHEAR ANGLE EQUATION

An experimental technique by which the shear angle of the metal cutting process can be measured involves a microscopic visual observation. Figure 2.25 illustrates an orthogonal cutting application where the cutting operation is reduced to a two-dimensional analysis. If a workpiece of the type shown in Fig. 2.25 is mounted on a milling machine table and a tool of the type shown is fixed to the columns of the miller, then a measurement of the shear angle can be made. A microscope with a hairline and an affiliated protractor can serve as the measuring instrument. By this technique, a series of experiments

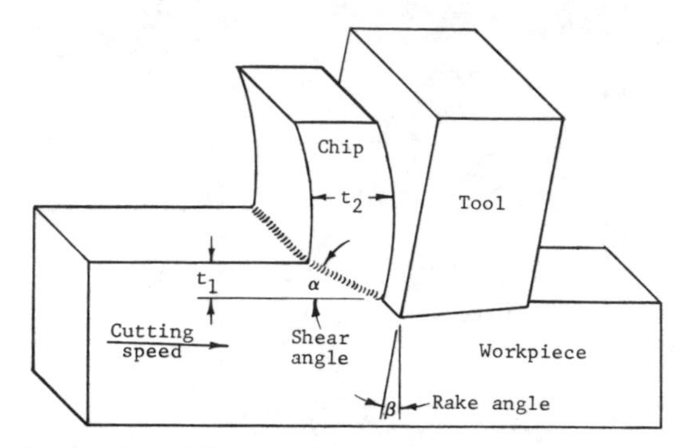

Fig. 2.25. Pictorial representation of orthogonal cutting.

can be carried out noting the effect on the shear angle of various cutting arrangements. As an example, the effect of the tool rake angle on the shear angle can readily be measured in this way.

Another technique by which the shear angle can be evaluated is by means of measuring the ratio of the feed (t_1) and the chip thickness (t_2). Figure 2.26 relates the chip thickness to the feed by means of two right triangles that share a common hypotenuse. This length signifies the common shear zone that serves as the demarcation line between the work material and the chip. It can be written as

$$\text{Shear length} = \frac{t_1}{\sin\alpha} = \frac{t_2}{\cos(\alpha - \beta)} \tag{2.20}$$

where t_1 is the feed, t_2 is the chip thickness, α is the shear angle, and β is the tool rake angle. In order to isolate the shear angle, the follow-

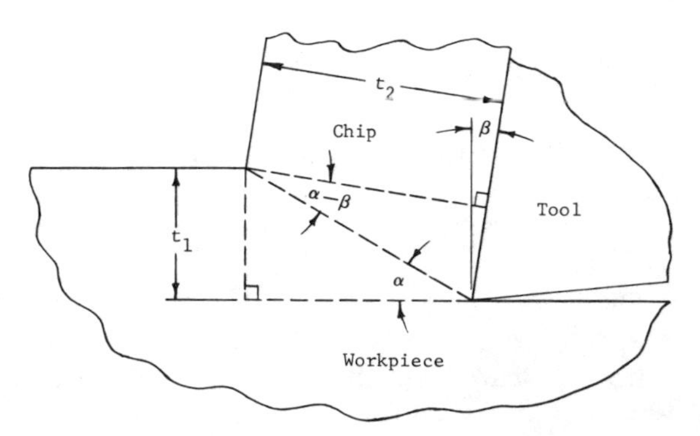

Fig. 2.26. Geometric relationship between feed and chip thickness.

ing algebraic manipulation is conducted. From Eq. 2.20,

$$\frac{t_1}{t_2} = \frac{\sin\alpha}{\cos(\alpha - \beta)}$$

Substituting the trigonometric identity

$$\cos(\alpha - \beta) = \cos\alpha\,\cos\beta + \sin\alpha\,\sin\beta$$

yields

$$\frac{t_1}{t_2} = \frac{\sin\alpha}{\cos\alpha\cos\beta + \sin\alpha\sin\beta}$$

or

$$\frac{t_1}{t_2} = \left(\frac{\cos\beta}{\tan\alpha} + \sin\beta\right)^{-1}$$

Isolating $\tan\alpha$ leads to

$$\frac{\cos\beta}{\tan\alpha} + \sin\beta = \frac{t_2}{t_1}$$

and

$$\frac{\cos\beta}{\tan\alpha} = \frac{t_2}{t_1} - \sin\beta$$

Finally,

$$\tan\alpha = \frac{\cos\beta}{t_2/t_1 - \sin\beta} \times \frac{t_1/t_2}{t_1/t_2}$$

or

$$\tan\alpha = \frac{(t_1/t_2)\cos\beta}{1 - (t_1/t_2)\sin\beta} \tag{2.21}$$

where t_1/t_2 = cutting ratio = r_a. Substituting the cutting ratio into Eq. 2.21 leaves

$$\tan\alpha = \frac{r_a\cos\beta}{1 - r_a\sin\beta} \tag{2.22}$$

As can be seen from Eq. 2.22, the tool rake angle has an influence upon the shear angle α. The wedging action of the tool is controlled by the rake angle. As the rake angle increases positively, an intuitive appraisal indicates that the cutting action will be somewhat eased, requiring less effort to cut the workpiece. As a result, the chip thickness should decrease. For this to happen, as an examination of the geometry of Fig. 2.26 reveals that the shear angle must increase in value. On the other hand, with less wedging action, that is, with the rake angle small or even negative, the cutting operation tends more toward a rough plowing action. This, in turn, leads toward a decrease in the shear angle, resulting in an increase in the chip thickness.

The shear angle is important because it can lead to a calculation of the area over which the shear force acts. This, in turn, enables a calculation to be made of the shearing force required to perform the cutting operation.

The shearing length is given in Eq. 2.20 as

$$L_s = \frac{t_1}{\sin\alpha} = \frac{t_2}{\cos(\alpha - \beta)}$$

In terms of the shear length, the shear area can be written as

$$A_{st} = L_s \times d \tag{2.23}$$

where A_{st} is the shear stress area (in.^{2}) L_s is the shear length (in.), and d is the depth of cut (in.). Figure 2.27 gives an example of the location of the shear area.

EXAMPLE 2.17. Determine the shear angle for a metal cutting operation that has a feed setting of 0.010 in./rev (0.254 mm/rev) and a tool rake angle of 5°. A measurement of the chip reveals a thickness of 0.043 in.

The cutting ratio is equal to

$$r_a = \frac{0.010}{0.043} = 0.2326$$

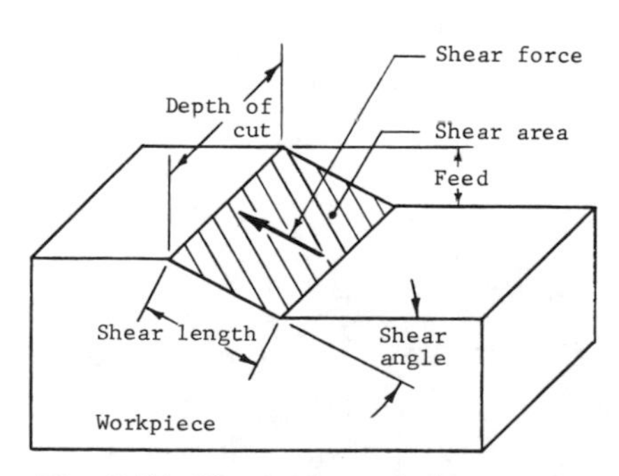

Fig. 2.27. Illustration of shear area.

Substituting numerical values in Eq. 2.22 yields

$$\tan\alpha = \frac{0.2326(0.9962)}{1 - 0.2326(0.0872)}$$

$$= 0.2365$$

Thus,

$$\alpha = 13.3^\circ$$

EXAMPLE 2.18. In an effort to reduce the cutting force, the tool rake angle in example 2.17 has been increased to 20°. With this rake angle, the chip thickness is found to be 0.025 in. (0.635 mm). From these data, calculate the shear angle.

The cutting ratio is equal to

$$r_a = \frac{0.010}{0.025} = 0.4$$

Substituting numerical values into Eq. 2.22 yields

$$\tan\alpha = \frac{0.4(0.9397)}{1 - 0.4(0.342)}$$

$$= 0.4354$$

Thus,

$$\alpha = 23.53^\circ$$

A graphical scale layout enables one to obtain a solution to examples 2.17 and 2.18 by a second method. Figure 2.28 lists the results. By drawing the feed, chip thickness, and rake angle to scale, a direct measurement of the shear angle can be made as shown in Fig. 2.28.

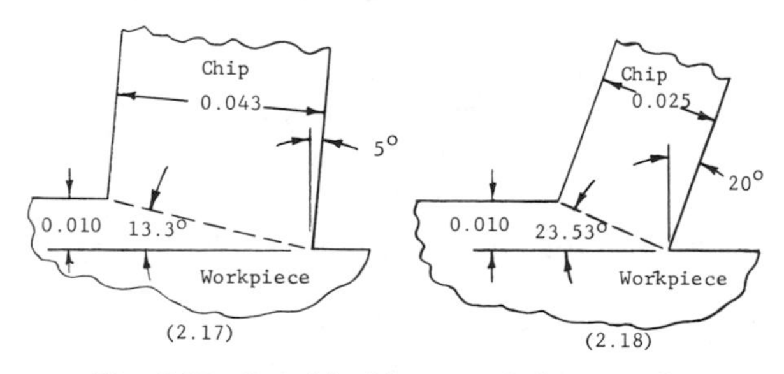

Fig. 2.28. Graphical layout of shear angle.

EXAMPLE 2.19. Determine the size of the shear area for examples 2.17 and 2.18 for a depth of cut setting of 0.250 in. (6.35 mm).

For example 2.17, the shear length can be calculated from Eq. 2.22 as

$$L_s = \frac{t_1}{\sin\alpha} = \frac{0.010}{\sin 13.3^\circ}$$
$$= 0.0435\,\text{in.}\ (1.05\,\text{mm})$$

and the shear area can be calculated from Eq. 2.23 as

$$A_{st} = L_x \times d$$
$$= 0.0435 \times 0.25$$
$$= 0.0109\,\text{in.}^2\ (7.03\,\text{mm}^2)$$

In a similar fashion for example 2.19,

$$L_s = \frac{0.010}{\sin 23.53^\circ}$$
$$= 0.02505\,\text{in.}\ (0.636\,\text{mm})$$

and

$$A_{st} = L_s \times d$$
$$= 0.02505 \times 0.250$$
$$= 0.00626\,\text{in.}^2\ (4.04\,\text{mm}^2)$$

EXAMPLE 2.20. If the assumption is made that the material being cut in examples 2.18 and 2.19 plastically flows in shear with a yield shear stress of 27,000 psi (186×10^6 Pa), calculate the percentage difference in shear force between the two examples.

The shear force for example 2.18 can be written as

$$\mathbf{F}_{s1} = S_s \times A_{st}$$
$$= 27{,}000 \times 0.0109$$
$$= 294.3\,\text{lb}\ (1309\,\text{N})$$

In a similar fashion for example 2.19,

$$\mathbf{F}_{s2} = S_x \times A_{st}$$
$$= 27{,}000 \times 0.00626$$
$$= 169.02\,\text{lb}\ (751.8\,\text{N})$$

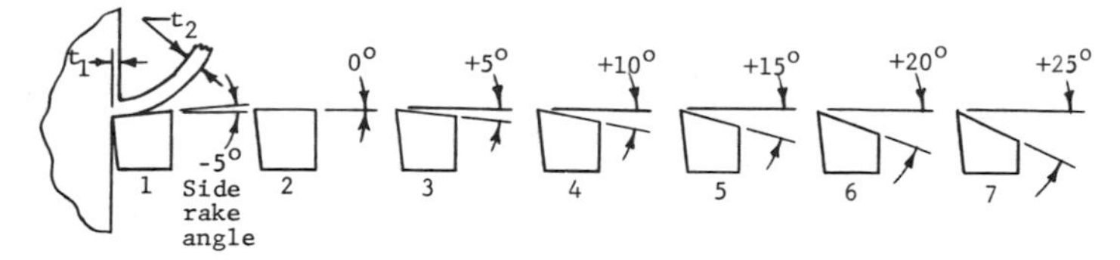

Fig. 2.29. Test tools for shear angle experiment.

As a result, the percentage decrease in shear force is equal to

$$\% \text{ decrease} = \frac{\mathbf{F}_{s1} - \mathbf{F}_{s2}}{\mathbf{F}_{s1}} \times 100$$

$$= \frac{294.3 - 169.02}{294.3} \times 100$$

$$= 42.6\%$$

2.9 SHEAR ANGLE EXPERIMENT

An experiment that can bring out the complexities of the metal cutting operation is one dealing with the evaluation of the shear angle as a function of the rake angle of the tool. This type of experiment can be conducted by a series of tests involving tools of differing rake angles. Figure 2.29 indicates the rake angles on tools used to turn on a lathe steel, SAE 1020, with a feed setting of 0.006 in./rev (0.152 mm/rev). The tool material used was high-speed steel. A compilation of the test results is given in Table 2.4. Placing the experimental data into Eq. 2.22

Table 2.4 Shear Angle Experimental Data

Test	Side Rake Angle, β (degrees)	Feed, t_1 [in. (mm)]	Chip Thickness, t_2 [in. (mm)]	Theoretical Shear Angle, α (degrees)
1	−5	0.006 (0.152)	0.0347 (0.881)	9.8
2	0	0.006 (0.152)	0.0370 (0.940)	9.23
3	+5	0.006 (0.152)	0.0415 (1.054)	8.2
4	+10	0.006 (0.152)	0.0233 (0.592)	15.0
5	+15	0.006 (0.152)	0.0200 (0.508)	16.7
6	+20	0.006 (0.152)	0.0210 (0.533)	15.9
7	+25	0.006 (0.152)	0.0164 (0.417)	20.0

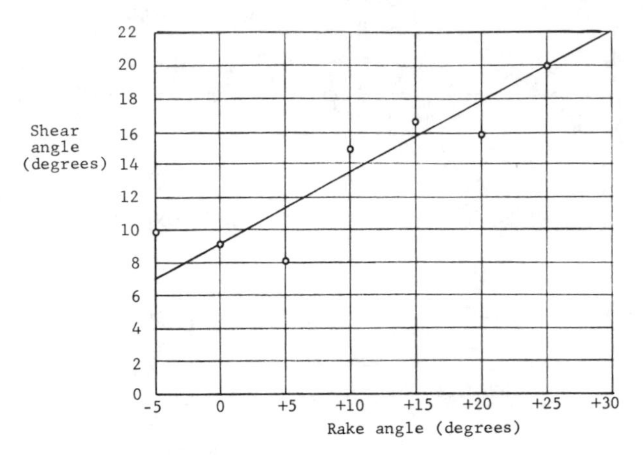

Fig. 2.30. Plot of shear experiment data.

in order to evaluate the theoretical shear angle reveals a wide scattering of points. Figure 2.30 illustrates graphically the calculated shear angle as a function of the rake angle. The data were collected by students as part of a tool engineering laboratory session dealing with the determination of the shear angle.[4]

Of interest is the wide scattering of the points on the graph. A straight line drawn through the second and last points in Fig. 2.30 indicates an approximation of the relationship between the shear angle and the rake angle for the tests conducted.

The wide scattering of the points is a testimony to the fact that the metal cutting process is a very complex phenomenon that is difficult to describe precisely especially with simple models. Nevertheless, by examining the test results, trend lines can be extracted and a conclusion can be reached that generally an increase in the side rake angle will produce a thinner chip as a consequence of an increase in the shear angle.

In examining the data in Table 2.4, it is noted that the shear angle exceeds the rake angle for rake angles less than 15°. This complies with the model used in Fig. 2.26. However, for the 20° and 25° rake angle tests, the calculated shear angle turned out to be less than the rake angle. Figure 2.31 gives a geometric representation of this case, which differs from Fig. 2.26 by a shift of the triangle representing the shear length and the chip thickness. Since β is greater than α in Fig. 2.31, the angle describing the shear length, known as the *chip thick-*

[4] Fryderyk E. Gorczyca, "Determination of Shear Angle," Tool Engineering Lecture Notes, Southeastern Massachusetts University, North Dartmouth, Massachusetts, 1985.

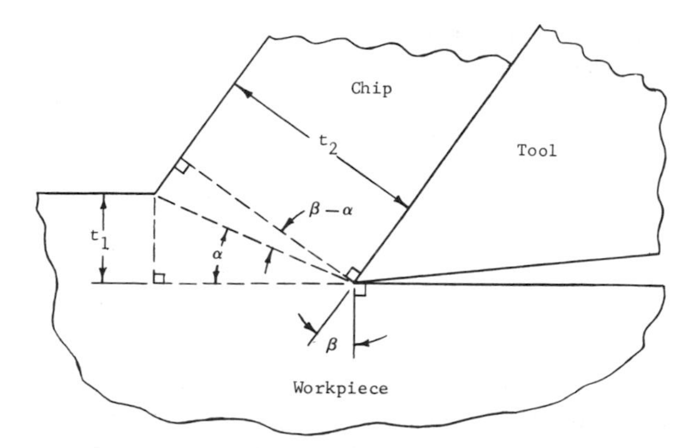

Fig. 2.31. Geometric relationship of feed and chip thickness for the case where the rake angle is greater than the shear angle.

ness triangle, has an interior angle of $\beta - \alpha$. An examination of the derivation of Eq. 2.22 indicates that it is valid for the model shown in Fig. 2.31 since

$$\cos(\alpha - \beta) = \cos(\beta - \alpha)$$

EXAMPLE 2.21. Figure 2.31 represents a scaled drawing of a metal cutting operation where $t_1 = 0.0215$ in. (0.546 mm), $t_2 = 0.051$ in. (1.295 mm), and $\beta = 35.50°$. From the given information, determine the shear angle first by graphical means and then by calculation.

Graphical measurement reveals that

$$\alpha = 24°$$

Substituting values into Eq. 2.22 yields

$$\tan\alpha = \frac{0.4216\cos 35.5°}{1 - 0.4216\sin 35.5°}$$

$$= 0.45446$$

Thus,

$$\alpha = 24.4°$$

2.10 CHIP FORMATION

The complexities of the interaction of the work material and the cutting tool during the metal cutting process evolve around the response of the material being cut to the forces and temperatures that are generated. The ideal case for a ductile material is represented by the shear angle models of Figs. 2.26 and 2.31. These models show a clean, narrow shear line where the plastic deformation takes place. Examinations of actual cutting operations indicate that the process is more complicated. The shear area is not a clean-cut line but rather a widely distributed zone. In addition, many times the properties of the work material do not lend themselves readily to a smooth flow in chip formation. As an example, for brittle materials such as cast iron, continuous shearing does not take place but rather rupture occurs intermittently, causing the chip to emerge in segments. Chip analysis can lead to a description of the conditions of cutting.

High forces and temperatures can produce a welding action between the tool and the work material, causing the chip to flow over this welded built-up edge. High temperatures can also influence the friction force between the tool and the work material. A blunt tool, that is, one with a low rake angle, produces a thicker chip as a result of a higher normal force, which in turn, produces a high friction force that tends to elevate the chip temperature as a result of the work performed. On the other hand, a more pointed tool, one with a high rake angle, has a tendency to reduce the chip thickness due to a reduction of the work being performed on the chip. A reduction in the magnitude of the forces necessary to form the chip ultimately leads to a lower chip temperature.

An examination of the shape of chips produced in the metal cutting process reveals a variety of configurations. These depend not only on the type of material being machined, but also on the conditions under which machining takes place. Figure 2.32 shows a segmented chip flow of a brittle material, where each segment undergoes a deformative pulselike cycle before it ruptures from the work material.

Chips have been classified into three different types: continuous, continuous with a built-up edge, and discontinuous.[5] Sample models of these types of chips are shown in Fig. 2.33. The least complicated type of chip is the continuous ribbonlike chip. It can usually be produced with a sharp, high-rake-angle tool machining a ductile material. In the event that there is no welding action between the tool and the work material being machined, the chip flows uniformly over the tool, producing a smooth ribbonlike continuous chip. This is illustrated in Fig. 2.33(a).

Complications develop when large forces combined with high heat cause a welding action to take place between the tool and the mate-

[5] Hans Ernst, "Physics of Metal Cutting," ASM paper on machining of metals, Cleveland, Ohio, 1938.

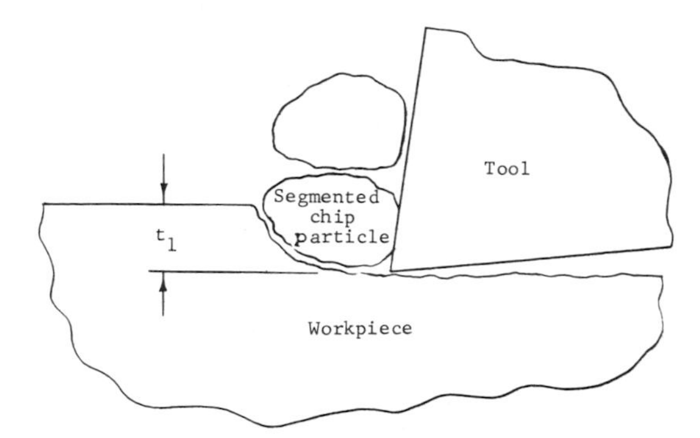

Fig. 2.32. Segmented chip from brittle material.

rial being machined. This condition is illustrated in Fig. 2.33(b). The welded portion of the chip deposited on the tool is harder than the remainder of the chip due to additional work hardening that has taken place. This built-up edge serves temporarily as the cutting surface, forcing the rest of the chip to flow over this deposit. As the built-up edge grows in size, it eventually is dislodged as a result of an increase of the forces acting on it. Chips of this type are called *chips with a built-up edge*. They can usually be produced with a ductile material and low-rake-angle tools running under conditions that produce sufficient heat and frictional forces to cause the welding action to occur between the tool and the chip.

The third classification of chip results in a situation where the material being machined is brittle. The tendency in this case is for the chips

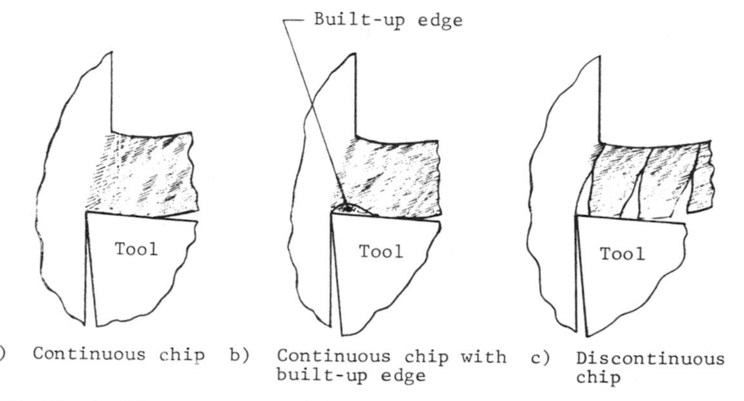

Fig. 2.33. Three different types of chips. (a) Continuous chip. (b) Continuous chip with built-up edge. (c) Discontinuous chip.

to be segmented. This is due to the property of limited deflection under shear. The results are a series of fractures causing a separation of the chip into short segments that may or may not adhere to each other, depending on the machining conditions. Chips of this type are called *discontinuous*. This is illustrated in Fig. 2.33(c).

2.11 A SUMMARY OF THE CUTTING PROCESS

The metal cutting process is a complex phenomenon. For purposes of developing a basic understanding, the metal cutting process has been reduced to a simplified single-point tool case. Machining operations such as turning, milling, shaping, and drilling have been shown reduced to a single-point tool case. Affiliated basic calculations are also listed for speed and cutting time evaluations. The chip-forming sequence has been considered in terms of a concentrated force acting on the grains being cut, placing them in a stressed state that causes plastic flow. This deformation leads to the formation of the chip, which ultimately is removed from the parent material.

An examination of the stress–strain diagram of a material allows one to gain an appreciation of the physical states through which a grain travels in the chip-forming process. In addition, the area under this curve gives a measure of the ductility of a material. To experimentally determine the stress–strain relationships of a material, tests involving loading a specimen can be conducted where corresponding deformations are recorded. These data, in turn, can be converted to graphs of the type shown in Fig. 2.16. A brittle material exhibits little deformation beyond the yield point, whereas a ductile material displays a large amount of deformation beyond the yield point. This is known as the *plastic deformation region*.

In the case where the material being cut is brittle and exhibits a rupture rather than a plastic flow, the chips formed are segmented or discontinuous. On the other hand, a ductile material readily responds to plastic deformation and produces a ribbonlike continuous chip.

The work that is performed in generating a chip by plastic flow is not recoverable and is converted to heat. If exact values of the forces and the corresponding deformations during the chip-forming operation are available, then the work performed can be calculated. A conversion of this work to a temperature rise in the chip, in turn, can be calculated through the use of the mechanical equivalent of heat.

A simple model of the chip-forming operation indicates that a large deformation of a standard segment of the parent material takes place in forming the chip. The shear angle clearly defines a line along which plastic deformation transpires. Of interest is the fact that the deformation lines have an angle greater than that of the shear angle. This is due to the flow of the material in forming the chip.

The shear angle can be observed experimentally through the use of a microscope. Calculations of the shear angle can also be determined through the use of a graphical layout as well as through a derived equation that takes into account the feed, chip thickness, and rake angle of the tool.

Experimental observation reveals that the metal cutting process is a complex procedure. Factors that affect the forces as well as the heat generated have a profound influence on the formation of the chip. An examination of different chips enables a general classification into three categories: continuous, continuous with a built-up edge, and discontinuous.

PROBLEMS

PROBLEM 2.1. Using Figs. 2.4 and 2.5 as a guide, determine the rpm setting as well as the volumetric rate of machining for the following conditions: cutting speed, 175 ft/min; diameter to be turned, 3 in.; depth of cut, 0.200 in.; and feed, 0.010 in./rev.

Answers: Rpm = 191 rpm
$R_v = 3.6\ \text{in.}^3/\text{min}$

PROBLEM 2.2. For a horizontal milling operation similar to that shown in Fig. 2.7(a), determine the following: (a) the rpm machine setting for an 88.9-mm-diameter cutter to run at 22.86 m/min; (b) the volumetric rate of machining for a milling feed setting of 44.5 mm/min, with a depth of cut of 6.35 mm and a width of cut of 38.1 mm; and (c) the feed per tooth if the cutter has 12 teeth.

Answers: (a) Rpm = 81.9 rpm
(b) $R_v = 0.179\ \text{cm}^3/\text{sec}$
(c) $F_t = 0.0457$ mm/tooth

PROBLEM 2.3. Determine the time required to machine a shaft similar to that shown in Fig. 2.9. The length to be cut is 5 in., with a feed of 0.008 in./rev and an rpm setting of 80 rpm. Assume a $\frac{1}{16}$-in. approach and a $\frac{1}{16}$-in. overtravel allowance.

Answer: $T_m = 8$ min

PROBLEM 2.4. (a) Calculate the offset milling cutter distance, similar to that shown in Fig. 2.10, for a 101.6-mm-diameter cutter that is to cut a slot 25.4 mm deep.

(b) Determine the time required to mill the slot of part (a) in a workpiece 254 mm long, with a milling machine feed rate setting of 12.7 mm/min. Assume 1.59-mm allowance for approach and 1.59-mm allowance for overtravel.

Answers: (a) $X = 44$ mm
(b) $T_m = 23.7$ min

PROBLEM 2.5. Determine the cutting speed for a turning operation similar to that shown in Fig. 2.4 if the machine is set for 75 rpm and the diameter of the workpiece is 3.5 in.

Answer: $V = 68.7$ ft/min

PROBLEM 2.6. In Fig. 2.11, the resultant force acting on the tool is conveniently resolved into two perpendicular components, the normal force and the friction force. These are shown as forces acting on the chip. If the normal force is equal to 1379 N and the friction force is equal to 756 N, determine the magnitude of the resultant force.

Answer: $\mathbf{R} = 1573$ N

PROBLEM 2.7. A test specimen of the type shown in Fig. 2.13 is loaded with a force of 7500 lb. The diameter of the specimen is 0.500 in. It is noted that the change in length as a result of the loading is 0.002548 in. over a gauge length of 2 in. It is further observed that upon release of the load, the specimen returns to its original gauge length of 2 in. without any permanent deformation. Determine the stress and strain of the material when it is loaded with the 7500-lb force. In addition, evaluate the modulus of elasticity of the material.

Answers: Stress = 38,216.6 psi
Strain = 0.001274 in./in.
$E = 30 \times 10^6$ psi

PROBLEM 2.8. The following data were compiled from a compression test on a 25.4-mm cube of metal:

Load (N)	Deflection (mm)
44,482	0.0462
66,723	0.0693
88,964	0.0924
106,757	0.1108
111,206	0.1668

From the data, determine the modulus of elasticity and the yield stress.

Answers: $E = 37.92 \times 10^9$ Pa
$S_{yp} = 172 \times 10^6$ Pa

PROBLEM 2.9. A structural steel member has a yield stress of 110×10^3 psi in tension. Estimate the range of the yield point for shear of this material.

Answers: $S_{syp} = 60.5 \times 10^3$ psi to 66×10^3 psi

PROBLEM 2.10. (a) How much work is performed on a steel cube as a result of shear forces acting on it, similar to that illustrated in Fig. 2.21(c), if an average yield force of 244.6×10^3 N permanently deforms the cube through a displacement of 9.398 mm?

(b) What would be the expected temperature rise of the cube if all the work performed during the plastic deformation was converted to heat and retained by the cube? Assume a 25.4 mm cube.

Answers: Work = 2296 N-m
$\Delta T = 70.25$°F

PROBLEM 2.11. Determine the amount of heat energy generated by a frictional force of 500 lb acting over a distance of 1 in.

Answer: $Q = 0.0535$ Btu

PROBLEM 2.12. It is desired to evaluate the amount of work that a member absorbed prior to fracture. An examination of the stress–strain diagram reveals that an estimate can be made that the member was exposed to an average yield force of 133,000 N during the period when it deformed a distance of 17.22 mm prior to fracture.

Answer: Work = 2290 N-m

PROBLEM 2.13. In an effort to specify the modulus of rigidity for an alloy, a shear test was conducted from which the following data were extracted:

Stress (psi)	Strain (in./in.)
10,000	0.00083
20,000	0.00167

From the given experimental data, determine the modulus of rigidity for the alloy.

Answer: $G = 12 \times 10^6$ psi

PROBLEM 2.14. Calculate the shear area for a metal cutting operation similar to that shown in Fig. 2.23, where the feed is 0.254 mm/rev, the depth of cut is 3.175 mm, and the shear angle is 20°.

Answer: Area = 2.355 mm^2

PROBLEM 2.15. It is desired to calculate the potential temperature rise in a steel alloy chip as a result of the shearing action of a machining operation similar to that represented in Fig. 2.23. The feed is 0.012 in./rev, the depth of cut is 0.1875 in., and the shear angle is measured to be 21°. An examination of the shear zone reveals an average width of 0.006 in., caused by an average yield shear stress of 30,000 psi that permanently deforms the shear zone element a distance of 0.011 in.

Answer: $\Delta T = 189.85°F$

PROBLEM 2.16. Given the model shown below, where a frictional force of 489.3 N acts over a contact distance of 1.5875 mm on a segment of chip of the dimension shown, calculate the rise in temperature of the steel chip due to the work of the frictional force. Assume that 80% of the heat generated is absorbed by the chip.

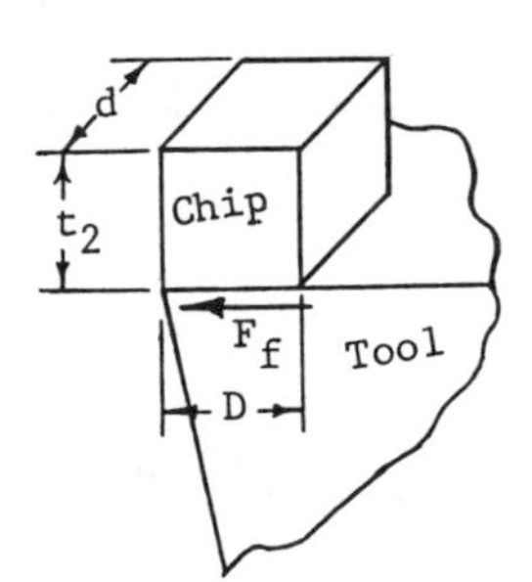

Note: **D** = 1.5875 mm
t_2 = 1.524 mm
d = 3.175 mm

Answer: $\Delta T = 40.52°F$.

PROBLEM 2.17. Calculate the length of the shear zone for the case where the feed is equal to 0.018 in./rev, the shear angle is 20°, and the rake angle is 10°.

Answer: $L_s = 0.0526$ in.

PROBLEM 2.18. Given the data of problem 2.17 ($L_s = 1.336$ mm), determine the thickness of chip for this machining operation.

Answer: $t_c = 1.316$ mm

PROBLEM 2.19. Determine the shear angle for the case where the feed is 0.020 in./rev, the rake angle is 8°, and the thickness of chip is 0.050 in.

Answer: $\alpha = 22.8°$

PROBLEM 2.20. What can be expected for a chip thickness for the case where the tool rake angle is 15°, the feed is 0.635 mm/rev, and the shear angle is 25°?

Answer: $t_2 = 1.478$ mm

PROBLEM 2.21. How much heat does a frictional force of 200 lb generate when it acts over a distance of 10 ft?

Answer: $Q = 2.57$ Btu

PROBLEM 2.22. What is the expected strain in a material that has 6.89×10^6 N-m/m^3 of work performed on it by a stress of 310×10^6 Pa?

Answer: $e = 0.0222$ m/m

PROBLEM 2.23. How much of the strain calculated in problem 2.22 takes place in the plastic region if the average yield stress on the material is 45,000 psi and the modulus of elasticity is 30×10^6 psi?

Answer: Strain plastic region = 0.0207 in./in.

PROBLEM 2.24. In generating a built-up edge on a tool, sheared material is deposited on an area of (1.778×2.54) mm^2. Calculate the equivalent friction force of this shearing deposit if the average shearing yield stress is 186×10^6 Pa.

Answer: $F_f = 840$ N

PROBLEM 2.25. The results of an experiment reveal that the effects of using a lubricant on a machining operation increases the cutting ratio by a factor of 1.4. If the tool rake angle is 10°, calculate the difference in the shear angle that takes place for a case where the feed is 0.015 in./rev and the original chip thickness was 0.045 in.

Answers: $(t_2)_2 = 0.032$ in.
$\Delta\alpha = 7.38°$

BIBLIOGRAPHY

Baumeister, Theodore, *Marks' Standard Handbook for Mechanical Engineers*, McGraw-Hill, New York, 1987.

Beer, Ferdinand P., and Johnston, E. Russell, Jr., *Vector Mechanics for Engineers*, McGraw-Hill, New York, 1977.

Collier's Encyclopedia, Macmillan, New York, 1984.

Counsell, Alden W., Test for True Stress and Strain, unpublished laboratory experiment, Northeastern University, Boston, 1949.

Ernst, Hans, "Physics of Metal Cutting," ASM paper on machining of metals, Cleveland, 1938.

Frazer, Charles G., *Half-Hours with Great Scientists*, Reinhold, New York, 1948.

Gorczyca, Fryderyk E., "Determination of Shear Angle," Tool Engineering Lecture Notes, Southeastern Massachusetts University, North Dartmouth, Massachusetts, 1985.

Lindberg, Roy A., *Processes and Materials of Manufacture*, Allyn and Bacon, Boston, 1977.

Materials in Design Engineering, Vol. 58, No. 5, 1963.

Muhlenbruch, Carl W., *Experimental Mechanics and Properties of Materials*, Van Nostrand, New York, 1955.

Oberg, E., Jones, F. D., and Horton, H., *Machinery's Handbook*, 22nd ed., Industrial Press, New York, 1984.

Sen, Gapal C., and Bhattacharyya, Amitabha, *Principles of Metal Cutting*, New Central Book Agency, Calcutta, India, 1969.

Timoshenko, S., *Strength of Materials*, Van Nostrand, New York, 1955.

Wilson, Frank W., *Fundamentals of Tool Design*, Prentice-Hall, Englewood Cliffs, New Jersey, 1962.

World Book Encyclopedia, Field Enterprises Educational Corp., Chicago, Illinois, 1970.

3 Cutting Tool Materials

3.1 INTRODUCTION

The materials from which metal cutting tools are made have a profound influence on the performance of the tool. In many cases, conditions under which tools function, in addition to the expected performance of the tool, dictate the materials from which they are made. This is especially true in applications where high production requirements place tools in operating states that produce an environment involving high stresses, elevated temperatures, and harsh abrasion. As a result, tools that are effective in performing the function of cutting metal must possess certain physical properties.

Listed among these are strength properties that enable the tool to resist high forces without yielding, as well as hardness characteristics that prevent failure through brittle fracture. A cutting tool material must be resilient enough to absorb shock without permanent deformation. High-temperature hardness qualities are also of importance. These thermal characteristics permit the tool to retain its strength and hardness under operating conditions that have high temperatures. Wear resistance is another important characteristic that a tool material must possess. This property allows the tool to perform under high-pressure and high-temperature conditions without appreciable erosion of the cutting surface.

Since the metal cutting tool operates under localized high temperatures, the retention of the cutting edge at elevated temperatures is crucial. A low frictional resistance (low coefficient of friction) of the tool surface is a desirable characteristic because it reduces the frictional force produced by the chip sliding over the tool. Since this work (frictional force × displacement) is converted to heat, a tool with a low coefficient of friction will operate at a lower temperature than one with a higher coefficient of friction.

In this chapter, the major tool materials, namely, carbon tool steels, high-speed steels, cast alloys, carbides, ceramics, and diamonds, will be discussed in light of their particular applications. The heat-treating operation is introduced with the carbon tool steels to describe how the hardness of steel is altered to advantageously change the physical properties of the material. Since cutting tools operate in what may be considered a hostile high-temperature environment, the retention of hardness (synonymous with retention of the cutting edge) as a function of temperature gives a measure of the effectiveness of the tool material. This characteristic is illustrated on a relative scale, lending an insight into the effectiveness of different tool materials at specific operating temperatures.

Since the contact between the tool and the work material is on the surface of the tool, tool surface treatments and coatings are used to improve the performance of the tool. These treatments are discussed in this chapter along with the application of cutting fluids. The effectiveness of cutting fluids on increasing the efficiency of tools is achieved by means of a twofold influence. The first of these is through a reduction of detrimental temperature and the second is through a reduction of friction forces. These reductions cause an increase in the life of the tool, enabling it to operate for a longer period of time.

Examples of comparative applications of different tool materials are given in this chapter. General relationships between tool life (time required to wear out a tool) and temperature are also introduced to point out the effectiveness of tool cooling techniques on production. The evaluation of how a cutting fluid changes the metal cutting process is introduced to show that what may on occasion seem to be a small change in an operation can have a profound influence on the end results.

Finally, in this chapter, Taylor's tool life equation is introduced. Through this equation, quantitative measures of production as well as operating cycles of different tool materials can be made. In turn, these measures can be applied to the solution of problems dealing with the comparison of the consequences of using different tool materials.

3.2 CARBON TOOL STEELS

Carbon is the influencing element in carbon tool steels. Its content in tool steels ranges from 0.6 to 1.4%. When the carbon combines with the iron, it forms a carbide that provides characteristics that are influenced by heat treatment, enabling control of physical properties such as hardness, strength, and wear resistance.

In hardening a carbon tool steel, it is heated to approximately 1500°F. During the heating, the normal and soft pearlite structure is changed to a solid solution called austenite. If the austenite is successfully cooled from its high temperature by a rapid quench (rapid cooling) until a temperature of approximately 200°F is reached, then another change takes place in the austenite. It is transformed into a structure known as martensite. This is the structure of fully hardened steel. The hardening process of carbon steel is a simple operation that transforms a soft and ductile steel to a hard and brittle steel.

However, after hardening, the steel cannot successfully be used because it is too brittle. To make the hardened steel usable, it must be further treated. This second heat-treating operation is called *tempering.* It involves reheating the hardened steel to some predetermined temperature for a certain length of time, to be followed by slow cooling back to room temperature. The tempering process serves to relieve the highly stressed state following hardening and tends to decrease the brittleness of the steel. This increases the toughness of the steel and allows it to be used successfully.

Carbon tool steels, when hardened and tempered, exhibit qualities such as high strength, high hardness, and wear resistance. In applications where slow speeds and low operating temperatures are encountered, carbon tool steels can be successfully used for metal cutting. Their great disadvantage lies with the characteristic that they have very poor hot hardness, that is, they do not retain hardness at elevated temperatures. This characteristic limits their effectiveness to applications involving (1) hand tools such as taps and files, (2) cutting tools used to machine soft materials such as brass, and (3) cutting tools that use low speeds, such as reamers and broaches. A typical designation for a high-carbon standard steel is given in Table 3.1.

After heat treatment, carbon tool steels retain a keen edge in applications where high heat and high abrasion are not present. Other advantages are that carbon tool steels are easily machined and can be heat treated to produce a hard surface with a tough core. Of special interest is a view of the operational range of carbon tool steels. Figure 3.1 is presented to show a general relationship of the hot hardness qualities of the three major cutting tool materials, that is, carbon tool steels, high-speed steels, and carbides. The dashed line in Fig. 3.1 represents a typical hardness of a workpiece at room temperature.

Table 3.1. Typical Designation for a High-Carbon Standard Steel

Steel Designation, SAE	Chemical Composition (%)			
	Carbon	Manganese	Phosphorus	Sulfur
1090	0.85–0.98	0.60–0.90	0.040	0.050

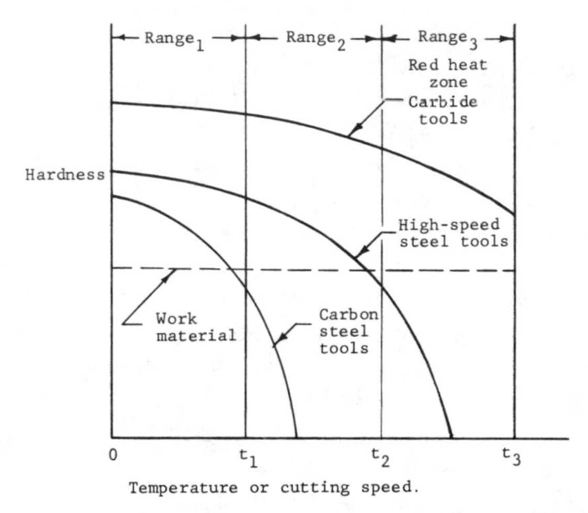

Fig. 31. Typical hardness–temperature relationship of carbon steel tools, high-speed steel tools, and carbide tools.

Since the hardness of the tool, for it to remain effective, must exceed that of the work material, it becomes obvious from Fig. 3.1 that carbon tool steel experiences a rapid decrease in hardness as the temperature rises. Because the temperature is directly related to cutting speed in a metal cutting operation, that is, the faster the cutting speed, the higher the temperature, it becomes obvious from Fig. 3.1 that tools made from carbon steels cannot function at high cutting speeds. Range 1 in Fig. 3.1 represents a low-speed range, and as can be seen the carbon steel will become ineffective toward the upper portion of this range.

As the demand for higher productivity dictated higher cutting speeds, which, in turn, produced correspondingly higher temperatures, the emergence of cutting tool materials to satisfy this requirement took place. The first of these materials with the distinct property of retention of hardness at elevated temperatures was high-speed tool steel.

3.3 HIGH-SPEED TOOL STEEL

High-speed steels are made from carbon steel with the addition of specified alloying elements such as molybdenum, chromium, tungsten, vanadium, and cobalt. These alloying elements provide especially

Table 3.2. Chemical Composition of Some High-Speed Tool Steels

	Chemical Composition (%)					
Grade	Carbon	Tungsten	Molybdenum	Chromium	Vanadium	Cobalt
T1	0.75	18.0		4.0	1.0	
T6	0.80	20.0		4.5	1.5	12.0
M1	0.80	1.5	8.0	4.0	1.0	
M6	0.80	4.0	5.0	4.0	1.5	12.0

desirable characteristics for cutting tool materials. Listed among these are: higher hardness; greater wear resistance; deeper hardness penetration; and perhaps most important, hot hardness, the ability to maintain a cutting edge at elevated temperatures. High-speed tool steels can maintain a cutting edge at temperatures (correlated with cutting speeds) that would very rapidly soften the cutting edges of carbon tool steels. For this reason, they have a distinct advantage over carbon tool steels. Figure 3.1 illustrates the hardness–temperature relative relationship of high-speed tool steels. As can be seen, high-speed tool steels can effectively be used in range 2, a temperature range where carbon tool steels quickly soften. The hardness of carbon tool steels declines rapidly above 400°F, whereas high-speed tool steels can maintain a cutting edge for long periods at temperatures as high as 1000°F. On a comparative scale, high-speed tool steels possess superior hot hardness and higher wear resistance when compared to carbon tool steels. A typical chemical composition for some high-speed tool steels is shown in Table 3.2. A common type of high-speed tool steel is grade T1, as shown in Table 3.2.[1] It contains 18% tungsten, 4% chromium, and 1% vanadium. In the literature, it is often listed with the symbol HSS-18-4-1.

The addition of different combinations of alloying elements affects the properties of high-speed tool steels. These desired properties include high wear resistance, retention of physical properties at elevated temperatures, high impact strength, high tensile strength, and high toughness. The right combination of these properties allows a tool to function under the adverse high temperature and high stress conditions present during the cutting process.

An investigation into dominance of different elements on the properties of high-speed tool steels reveals that carbon is the hardening element in carbon tool steel, whereas tungsten provides this influence in high-speed tool steel. It provides for the formation of a desirable

[1] *Machining Data Handbook*, Metcut Research Associates, Cincinnati, Ohio, 1972, p. 969.

dense and fine structure. Tungsten increases hardenability as well as the strength and toughness at high temperatures. This imparts the very desirable property of red hardness. Another element that is used in high-speed tool steel is molybdenum. It serves in a fashion similar to that of tungsten, providing strength and hardness at elevated temperatures. Chromium is also used as an alloying element in high-speed tool steel. It has the effect of hardening and toughening the steel. Other influences of chromium are increases in wear and abrasion resistance as well as increases in the resistance to fatigue failure and corrosion.

As can be seen in Table 3.2, vanadium is also used as an alloying element in high-speed tool steel. When added to carbon steels, it has a strong affinity for carbon, resulting in the formation of vanadium carbide. This formation provides for high wear resistance, which, along with hot hardness, is a general characteristic of high-vanadium high-speed tool steel.

Cobalt is another element that is found in high-speed tool steel. High-speed tool steels with cobalt possess the property of being able to be hardened to high levels. These high-hardness cobalt steels can be treated to hardness values in the high ranges. In general, high-speed tool steels with cobalt are less tough than conventional high-speed tool steels but do exhibit greater hot hardness and greater wear resistance

As we climb the scale of materials with greater hot hardness, we come across cast cobalt tools. These have applications at temperature ranges and cutting speeds between high-speed tool steels and cemented carbide tools.

3.4 CAST COBALT ALLOY TOOLS

Alloy systems consisting of cobalt, chromium, tungsten, and carbon display high hardness and toughness and are used as tool materials. These tools are cast and are finished by grinding to the desired shape. Cast cobalt tools are made by melting the particular analysis. The melt is poured into a mold and allowed to solidify to a specific shape, with a small allowance for finish grinding. With this technique, the tools are at maximum hardness when cooled in the molds. They cannot be hot or cold worked because of their brittleness, and they do not respond to heat treatment.

Figure 3.2 illustrates the relative hardness–temperature relationship of cobalt alloys, high-speed steels, and ceramics. As can be seen, the cast cobalt-base alloys display greater retention of hardness at

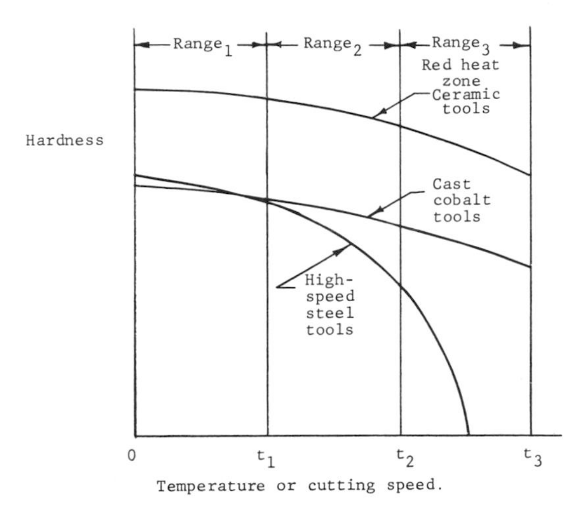

Fig. 3.2. Typical hardness–temperature relationship of high-speed steel tools, cobalt tools, and ceramic tools.

elevated temperatures than the high-speed steels, although they have a lower hardness at lower temperatures. In addition, cobalt-base tools possess relatively lower coefficients of surface friction, resulting in the generation of lower temperatures as a result of sliding contact between the chip and the tool.

As can be seen in Fig. 3.2, cast cobalt tools occupy a range shared with high-speed tools in range-1 temperatures. At higher temperatures, cast cobalt tools fall in a hardness range between high-speed tools and carbide tools. Depending on the particular application, they can be used successfully. The *Manual on Cutting of Metal*[2] states that for relatively light cuts, particularly finishing and forming cuts, where the temperature at the nose of the tool is less than 1100°F, high-speed steel will usually prove superior. For higher temperatures, the cobalt-base metals will often be more satisfactory and may therefore be used at higher cutting speeds.

Usually with increases in hardness, there is a corresponding increase in brittleness, resulting in a tendency for the tool to chip. As we move up the hardness scale with different tool materials, there is a lower resistance to shock, which can lead to failure by chipping of the cutting edge. As a result, with the more brittle tool materials, excessive chip thickness, chatter, and interrupted cuts should be avoided.

[2] *Manual on Cutting of Metal,* American Society of Mechanical Engineers, New York, 1952, p. 41

Table 3.3. Typical Chemical Composition of Two Cobalt-Base Alloys

Tool Material	Composition Range (%)				
	Cobalt	Chromium	Tungsten	Carbon	Other
Stellite 19	55	30	12	1.7	1.3
Tantung 58	42	25	25	3.0	5

Table 3.3 lists two examples of the chemical composition of cast cobalt tools.[3]

3.5 CEMENTED CARBIDE TOOLS

Cemented carbide tools form a widespread tool material family of hard carbide composition. These materials are produced by methods of powder metallurgy and are composed of the carbides of tungsten, titanium, or tantalum sintered or cemented in a matrix binder, which usually is cobalt. The combination of the basic constituents produces a metal that is characterized by its high hardness, high compression strength, and high hot hardness. As a result of these physical properties, sintered carbide tools exhibit exceptional tool performance in cases where other tools possessing lower physical properties fail rapidly. There are many grades for carbide tools. Usually the application recommendations are given in terms of such operations as roughing, finishing, and precision finishing. Affiliated characteristics of carbide tools further classify them with respect to shock resistance, wear resistance, and cutting temperature resistance.

Most carbide tool grades are listed under an industry code using the letter C followed by a numeral. In many cases, a particular grade can be found to perform a specific metal cutting operation. The application of carbide tools is wide. They are used to machine a large number of different types of work materials.

3.6 CERAMIC TOOLS

Ceramic tools are made primarily from aluminum oxides and are manufactured by power-metallurgical processes. These tools consist

[3] Haldon J. Swinehart, *Cutting Tool Material Selection*, American Society of Tool and Manufacturing Engineers, Dearborn, Michigan, 1968, p. 66.

primarily of very small particles of aluminum oxide (Al_2O_3) that are bonded together by a high-temperature sintering process. Additives such as chromium oxide, magnesium oxide, titanium oxide, nickel oxide, and refractory metal carbides are added to achieve favorable properties in the cutting edge. Ceramics, amongst tool materials, exhibit the best resistance to failure at elevated temperatures and usually perform better at high-speed ranges when compared to carbide tools. They also are very hard and possess high wear-resistant characteristics. Applications involve cutting conditions at elevated temperatures that would cause failure to other tool materials. As can be seen in Fig. 3.2, ceramic tools retain a high level of hardness at elevated temperatures. Since they retain a higher hardness at elevated temperatures than carbide tools, the expectation is that they will perform better in the high-speed ranges designated as the hot heat range in Fig. 3.2.

Disadvantages of the material as a tool revolve around its relatively low transverse rupture strength as well as its high brittleness, which limit its capabilities, especially in cases of interrupted cuts. Because of high brittleness, ceramic tools are susceptible to chipping, cracking, fracturing, and gradual wear by abrasion. In addition, ceramic tools exhibit a tendency toward a welding action between the tool and work material as well as a certain degree of plastic deformation of the tool at elevated temperatures. Figure 3.3 illustrates the relative relationship between hot hardness or wear resistance and strength or toughness for the four major cutting tool materials.

On a relative scale, aluminum oxide ceramic tools possess a combination of a very high hardness and a low toughness when compared to other cutting tool materials, as can be seen in Fig. 3.3. The major

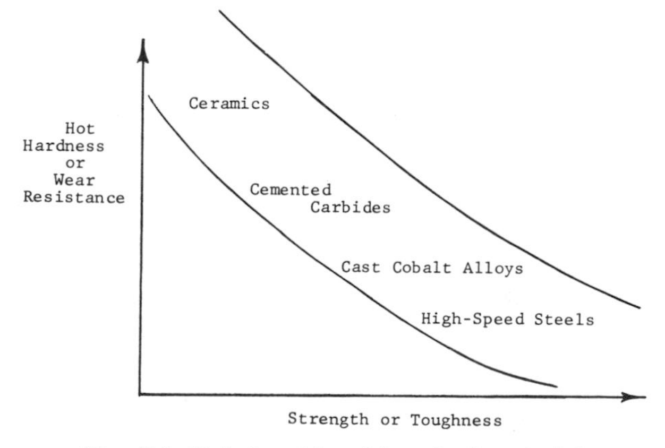

Fig. 3.3. Relationship of four tool materials.

application of ceramic tools is in high-speed turning of cast irons and very high-strength steels as well as in finishing operations at high-speed ranges.

3.7 DIAMOND TOOLS

Diamond is the hardest known natural material. It consists of the cubic crystalline form of carbon formed under conditions of tremendous heat and pressure. Natural stone tools as well as synthetic diamond tools in the form of sintered polycrystalline diamond powders are most suited to machine nonferrous metals such as copper and aluminum and also nonmetallic materials such as ceramics, carbon, and fiberglass composites. On the Knoop hardness scale, diamond (7800 kg/mm^2) is five times as hard as tungsten carbide (1600 kg/mm^2). Under proper conditions, that is, very fine cuts on readily machinable materials where high standards for dimensional accuracy and surface finish are demanded, diamond tools can machine very large numbers of parts without being resharpened. The main advantage of a diamond tool is that it possesses the ability to maintain accurate size and a smooth finish over a long production run while operating at high cutting speeds. They also have an advantage over other tools on work materials that have an abrasive effect on the cutting edge.

Diamond tools cannot be used for machining ferrous materials in cases that generate high temperatures. Under these conditions diamond tool materials disintegrate as they do in interrupted or roughing types of cuts. As a result, it is recommended to use diamond tools under conditions of minimum vibration or chatter. Impact cutting or interrupted cutting proves detrimental to diamond tools, as do inclusions in the work material such as hard-particle concentrations. The vulnerability of diamond tools to chipping due to its high brittleness mandates light feeds and continuous cuts for successful operations. High speeds are possible with diamond tools because of very low frictional forces as well as high heat conductivity tool characteristics that tend to provide relatively low operating temperatures at the cutting edge.

Heat is the main cause of failure of all cutting tools, including diamond tools. Chipping or burning can result from excessive heat. Economically, diamond tools applied under cutting conditions involving light cuts or nonferrous materials have been found to retain a cutting edge over a long production range, thus enabling a close tolerance to be held and scrap to be reduced. Although diamond tools may have a substantially higher initial cost than other tool materials, under specific operations where they produce at higher rates, generate less down time, and reduce scrap percentages, they can prove to be more economical to use.

3.8 SURFACE TREATMENT OF TOOLS

Surface treatments and coatings can improve the performance of a cutting tool. The objectives of treating the surface or placing a coating on the surface of a cutting tool center on adding favorable characteristics to the tool. Generally, the methods include polishing the surface, lubricating the surface, hardening the surface, and plating the surface. The results are usually increased wear resistance and reduced frictional surface forces that act between the tool and work material.

For high-speed steel tools, surface treatments such as nitriding, carburizing, coating, and oxidizing have been used to increase the integrity of the tool material. Nitriding has the tendency to increase hardness and abrasion resistance through an increase of nitrides in the surface layer of the tool. The heat treating provides a nitrided case over hardened alloy steel that ultimately leads to an increase in the life of the tool. Carburization has an effect similar to nitriding. It involves the process through which the surface layer of the tool receives an increased coating of carbides. It is a method of increasing the carbon content of the surface of steel by the process of absorption and diffusion. The process involves heating the steel while in contact with an appropriate carbonaceous medium for a period of time that enables absorption and diffusion of carbon to a desired depth of the tool. Upon heat treatment, a hard case of carbides is formed over the tool.

Coatings such as chromium, tungsten, and other alloys deposited on tool surfaces serve the function of lowering the coefficient of friction, increasing the abrasion resistance, and providing greater anti-chip welding characteristics. Techniques through which platings are deposited include electroplating, chemical deposition, and vapor deposition. Vapor-deposited coatings of titanium carbide and titanium nitride have been successfully used on cemented tungsten carbide inserts. As an example, a 0.0002–0.0003-in. layer of titanium carbide bonded to a tungsten carbide tool has had effects of reducing cutting forces by as much as 25% and temperatures by as much as 150°F when compared to a conventional cemented carbide tool.[4] These reductions

[4] Daniel B. Dallas, *Tool and Manufacturing Engineers Handbook*, McGraw-Hill, New York, 1976, p. 1-22.

are credited to the lowering of the coefficient of friction caused by the surface formed by the coating.

Surface oxide coatings placed on cutting tools have had beneficial effects on tool performance. The oxide coatings serve as a lubricating agent that increases the wetting ability of tools, thus increasing the ability to retain a cutting lubricant between the tool and the work material. These coatings, which are usually less than 0.001 in. thick, serve to decrease friction and enable the tool to operate at a lower temperature. In addition, the thin coating serves as a barrier for heat flow from the chip to the tool. Welding of the work material to the tool is also reduced through the deposit of oxide coatings.

3.9 CUTTING FLUIDS

The application of cutting fluids enhances the metal cutting process by cooling the tool and by providing lubrication to reduce friction. The relationship between tool performance (tool life) and heat is a very strong one. Wear experiments have revealed that tool wear increases exponentially with temperature and that tool life varies inversely as the nth power of the cutting temperature.[5] The exponent has a wide range, depending on the conditions of the metal cutting operation. Tool life is extremely sensitive to the cutting temperature. As a result, small reductions in cutting temperatures can produce large increases in allowable cutting speeds as well as large increases in the tool life of cutting tools.

Two results come into play when a cutting fluid is used. The first of these is that the cutting temperatures are reduced by the cooling effect of energy dissipation of the heat generated in the metal cutting process. Secondly, the cutting temperatures are reduced by the lubricating effect of the cutting fluid through friction reduction between the chip and the tool. As the friction force is reduced, so is the heat generated from the work that the friction force performs.

A simple experiment can be conducted to evaluate the effect of a cutting fluid on a metal cutting process. Figure 3.4 illustrates the relationship between the shear angle and the chip thickness. Shear angle (α_1) and the associated chip thickness (t_{c1}) represent cutting conditions without the use of a cutting fluid, that is, the operation is conducted dry. With the application of a cutting fluid, friction is reduced on the

[5] Daniel B. Dallas, *Tool and Manufacturing Engineers Handbook*, McGraw-Hill, New York, 1976, p. 2-2.

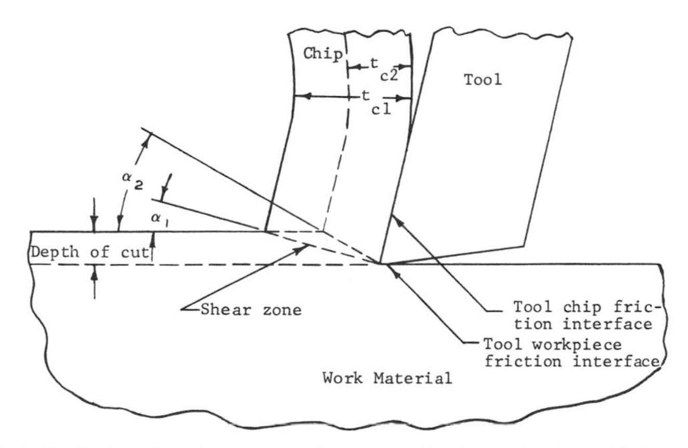

Fig. 3.4. Relationship between shear angle (α) and chip thickness (t_c).

tool–chip interface as well as on the tool–workpiece interface. The reduced friction acts to diminish the resistance that the work material offers to the cutting process and leads to an increase in the shear angle to a higher value (α_2). This, in turn, leads to a reduced shear zone that has a correspondingly smaller shear area. As a result, the chip thickness is reduced to t_{c2}. Since less shearing deformation takes place with shear angle α_2, less heat is generated in the formation of chip thickness t_{c2}. Less heat is generated because the reduced friction forces do less work.

A metal cutting operation changes as a result of reductions of frictional forces on the tool–chip interface as well as on the tool–workpiece interface. These reductions are credited to boundary lubrication, where additives in the fluid react chemically with the metal to form compounds on the metal surface. These compounds provide the mechanism for the friction reduction. They inhibit welding of the chip to the tool and provide a sliding layer between the chip and the tool. Of importance also is the heat-absorbing capability of the cutting fluid. With lower operating temperatures due to the cooling effect of the cutting fluid, the life of the tool is extended.

3.10 EXAMPLES OF TOOL APPLICATIONS

The performance of a cutting tool can be measured in terms of its useful life. In general, tool life can be defined as a time during which a tool satisfactorily functions. Oftentimes, this is measured in terms of

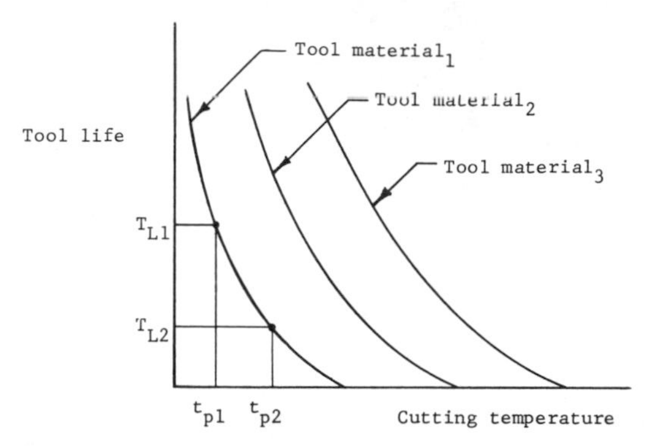

Fig. 3.5. Tool-life–cutting-temperature relationships for three different tool materials.

the time a tool is in use between successive sharpenings or the time a tool is in use before replacement is necessary to maintain satisfactory performance. Figure 3.5 illustrates the general relationship between tool life and cutting temperature for three different types of tool materials. As can be seen, each of these tool materials occupies a different range on the temperature scale.

The burnout temperature is represented by the intersection of the curves with the abscissa of the graph. This is the temperature at which the tool cannot effectively perform the cutting process. As the temperature is decreased from the burnout temperature, the tool life increases dramatically. The slope of the tool-life–cutting-temperature curve is high, signifying that large changes in tool life can be attained with relatively small changes in cutting temperature. This underscores the effectiveness of the cooling action of a cutting fluid on increasing tool life. Reductions in cutting temperatures can produce significant changes in tool life.

Tool wear increases as a function of the cutting temperature. Since tool life is inversely proportional to tool wear, an empirical expression can be written relating tool life and cutting temperature. Experiments designed to relate tool life and temperature have shown that for certain ranges of cutting temperature, the following expression can be written:[6]

$$t_p(T_L)^{1/n} = B \tag{3.1}$$

If both sides of the equation are raised to the nth power, the tool life

[6] Milton C. Shaw, *Metal Cutting Principles*, M.I.T. Press, Cambridge, Massachusetts, 1968, p. 12-7.

can be written as

$$[t_p(T_L)^{1/n}]^n = B^n$$

or

$$T_L = \frac{A}{(t_p)^n} \tag{3.2}$$

where T_L is the tool life (min), t_p is the cutting temperature (°F), A and B are constants of proportionality, and n is an exponent. Of interest is that the two constants in Eq. 3.2 can be evaluated if two experimental points involving cutting temperature and tool life are available for the range where the expression is valid. Figure 3.5 illustrates the two experimental points.

EXAMPLE 3.1. Determine the constants of Eq. 3.2 from the data compiled from the following two tests:

Test 1: $t_{p1} = 1023°\text{F}, \quad T_{L1} = 40 \text{ min}$

Test 2: $t_{p2} = 1100°\text{F}, \quad T_{L2} = 10 \text{ min}$

Substituting values in Eq. 3.2 yields

$$40 = A(1023)^{-n}$$

$$10 = A(1100)^{-n}$$

Isolating A gives

$$A = 40(1023)^n = 10(1100)^n$$

or

$$4 = \left(\frac{1100}{1023}\right)^n = (1.07527)^n$$

Solving for n yields

$$\ln 4 = n \ln(1.07527)$$

$$1.3863 = n(0.07257)$$

$$n = 19.1$$

In turn, A can be determined by substituting the value of n into Eq. 3.2. Therefore,

$$A = 40(1023)^{19.1}$$

or

$$= 1.2322 \times 10^{59}$$

As a result, for the example given, the relationship between tool life and cutting temperature is

$$T_L = 1.2322 \times 10^{59} \times (t_p)^{-19.1} \tag{3.3}$$

Of interest is that Eq. 3.3 documents the sensitivity of the tool life to small changes in cutting temperature. As can be seen from example 3.1, a 77°F decrease in temperature from 1100°F will, according to Eq. 3.3, increase the tool life by a factor of 4, namely, from 10 min to 40 min. This underscores the effectiveness of the cooling action of a cutting fluid on increasing tool life. The conclusion can be drawn that reductions in cutting temperatures can produce significant changes in tool life.

If Eq. 3.3 is written in the form of Eq. 3.1, it can be analyzed in terms of the temperature of 1-min tool life, that is, if T_L is set to 1 min, then B is equal to the temperature for the 1-min tool life. Raising both sides of Eq. 3.3 to the (1/19.1) power gives

$$(T_L)^{1/19.1} = (1.2322)^{1/19.1}(10^{59})^{1/19.1}[(t_p)^{-19.1}]^{1/19.1}$$

$$(T_L)^{0.05236} = (1.011 \times 10^{3.089}) \times (t_p)^{-1}$$

$$t_p(T_L)^{0.05236} = 1240 \tag{3.3a}$$

For the case where $T_L = 1$ min in Eq. 3.3a, we have $(T_L)^{0.05236} = 1$ and $t_p = 1240°$.

An important consideration in determining a particular tool material for an operation is the time required to perform the operation, which obviously is affiliated with the cost of the operation. An analysis of this type takes into account the recommended cutting speeds and feeds for the use of the different tool materials. The following example illustrates how the choice of a particular tool material can affect the time required to machine a given part.

EXAMPLE 3.2. Assume that a 2-in. (50.8-mm) diameter is to be turned in a lathe for a length of 10 in. (254 mm). Three tool materials are being considered for use: tool material 1 has a recommended cutting speed of 75 ft/min (22.86 m/min) and a feed of 0.015 in./rev (0.381 mm/rev); tool material 2 has a recommended cutting speed of 90 ft/min (27.43 m/min) and a feed of 0.015 in./rev (0.381 mm/rev); and tool material 3 has a recommended cutting speed of 150 ft/min (45.72 m/min) and a feed of 0.015 in./rev (0.381 mm/rev). Calculate the time required to machine the part with each of the tool materials.

The time required to machine a length of 10 in. can be written in terms of Eq. 2.5 as

$$T_m = \frac{L\pi D}{12fV}$$

A_p and O_t are considered negligible; from Eq. 2.1,

$$\text{Rpm} = \frac{12V}{\pi D}$$

The corresponding times then can be written as

$$T_{\text{tool 1}} = \frac{10 \times 3.14 \times 2}{12 \times 0.015 \times 75} = 4.65 \text{ min}$$

$$T_{\text{tool 2}} = \frac{10 \times 3.14 \times 2}{12 \times 0.015 \times 90} = 3.88 \text{ min}$$

$$T_{\text{tool 3}} = \frac{10 \times 3.14 \times 2}{12 \times 0.015 \times 150} = 2.326 \text{ min}$$

As can be seen from the corresponding machining times in example 3.2, the choice of a particular tool material can have a profound effect on the time required for a particular operation. As the example illustrates, tool 3 performs the turning operation in half the time required by tool 1 since it can function at twice the cutting speed of tool 1.

EXAMPLE 3.3. Given Eq. 3.3 as the relationship between tool life and cutting temperature, determine the operating temperature for a tool life of 20 min.

Substituting into Eq. 3.3 yields

$$20 = (1.2322 \times 10^{59}) \times (t_p)^{-19.1}$$

$$20(t_p)^{19.1} = 1.2322 \times 10^{59}$$

$$\ln 20 + 19.1 \ln t_p = \ln 1.2322 + 59 \ln 10$$

$$2.9957 + 19.1 \ln t_p = 0.2088 + 59(2.3026)$$

$$19.1 \ln t_p = 133.0665$$

$$\ln t_p = 6.9668$$

$$t_p = 1061^\circ\text{F}$$

Another illustrated example of special interest is the evaluation of the effect of a cutting fluid through the observation of the chip thickness. Figure 3.6 illustrates the shear area of a cutting process. The

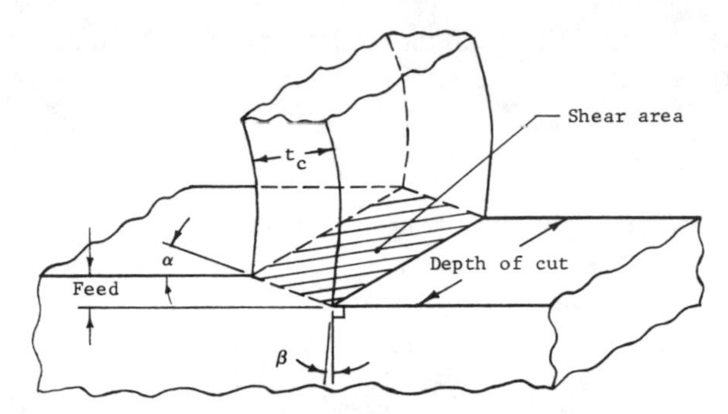

Fig. 3.6. Illustration of shear area in cutting process.

following example shows how a calculation through a determination of the change in the shear area can reflect the influence of the application of a cutting fluid on the cutting process.

EXAMPLE 3.4. An analysis of the influence of a cutting fluid is to be made on a metal cutting operation. The objective is to find the percent reduction of the shear area by measuring the chip thickness with and without the use of a cutting fluid. The feed is set at 0.010 in./rev (0.254 mm/rev) and the depth of cut is set at 0.100 in. (2.54 mm). The tool has a rake angle of 10°. The measurements from the experiment are as follows:

Test 1: Without cutting fluid

Chip thickness = 0.032 in. (0.813 mm)

Test 2: With cutting fluid

Chip thickness = 0.020 in. (0.508 mm)

The shear area represented by the shaded rectangular plane in Fig. 3.6 can be expressed as follows:

$$\text{Shear area} = \frac{\text{Feed}}{\sin(\text{shear angle})} \times \text{Depth of cut}$$

or

$$A_s = \frac{f}{\sin \alpha} \times d$$

Solving for the shear area machined dry yields

$$A_{\mathbf{sd}} = \frac{0.010}{0.3094} \times 0.100$$

$$A_{\mathbf{sd}} = 0.00323 \text{ in.}^2 \ (2.084 \text{ mm}^2)$$

where, from Eq. 2.22,

$$\tan \alpha_1 = \frac{0.3125(0.9848)}{1 - 0.3125(0.1736)} = 0.3254$$

$$\alpha_1 = 18.025^\circ$$

The shear area obtained while using the cutting fluid, written in terms of the thinner chip, is

$$A_{sf} = \frac{0.010}{0.4746} \times 0.100$$

$$= 0.00211 \text{ in.}^2 \ (1.361 \text{ mm}^2)$$

where, from Eq. 2.22,

$$\tan \alpha_2 = \frac{0.5(0.9848)}{1 - 0.5(0.1736)} = 0.5392$$

$$\alpha_2 = 28.33^\circ$$

Comparing the shear areas with and without a cutting fluid application reveals that the reduction in shear area is

$$\% \text{ reduction} = \left(\frac{A_{sd} - A_{sf}}{A_{sd}}\right) \times 100$$

$$= \left(\frac{0.00323 - 0.00211}{0.00323}\right) \times 100$$

$$= 34.67\%$$

As can be seen in example 3.4, the reduction in the shear area as a result of the application of a cutting fluid can be dramatic. If the assumption is made that the shearing force required to form the chip is proportional to the shearing area, then the conclusion can be reached that this is a corresponding reduction in the shearing force for the operation.

Another application of interest with respect to different tool materials deals with the tool life of a cutting tool as related to the cutting speed at which the tool is operating. Some of the earlier work relating cutting speed and tool life is credited to Frederick W. Taylor, who proposed a relationship of the form[7]

$$VT^n = C \tag{3.4}$$

[7] F. W. Taylor, *On the Art of Cutting Metals*, American Society of Mechanical Engineers, New York, 1906, p. 159.

where V is the cutting speed (ft/min); T is the tool life (min); C is a constant, dependent on cutting conditions; and n is the exponential constant, dependent on cutting conditions. In the literature, Eq. 3.4 is oftentimes referred to as *Taylor's tool life equation*. It expresses the relationship between the cutting speed and tool life in terms of two constants (n and C) that are functions of the cutting conditions involving a particular tool material machining a specific work material. Tool life (in minutes) can be measured in a variety of ways. In general, tool life is the length of time during which the tool performs satisfactorily. A convenient measure of this period is in terms of the time it takes to wear a tool to a particular wear limit beyond which the tool will no longer function satisfactorily. Let us assume that this is the criterion used in measuring the tool life. Figure 3.7 illustrates the relationship between tool life and cutting speed for different tool materials, plotted on a log–log scale.

Exponential equations of the form of Eq. 3.4 plot as straight lines on a log–log scale graph. The straight line is convenient for analysis insofar as the constants of Eq. 3.4 can then be directly evaluated. As can be seen in Fig. 3.7, constant n represents the slope of the straight line, whereas the constant C represents the y-intercept of the straight line. An examination of Eq. 3.4 reveals that if $T = 1$, then $T^n = 1$ and $V = C$. In other words, according to Eq. 3.4, when the tool life is equal to 1 min, then the cutting speed is equal to the constant C.

If the assumption is made that the cutting speed and tool life have the relationship as described in Taylor's equation, two tests can be conducted that will enable an evaluation of the constants n and C. Unfortunately, not all tool materials and work materials are so accom-

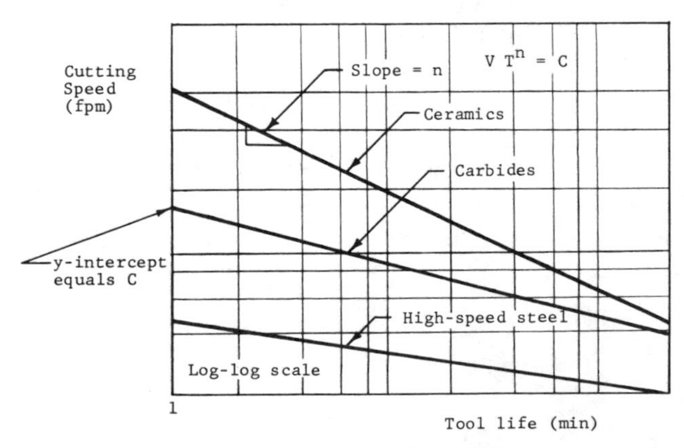

Fig. 3.7. Tool life versus cutting speed for different tool materials.

modating as to obey the strict relationship dictated by Eq. 3.4. Nevertheless, in certain restrictive regions of cutting speed, Eq. 3.4. can be verified by conducting a test.

EXAMPLE 3.5. Two tests are conducted to evaluate the constants n and C of Eq. 3.4 using a carbide tool machining a mild steel. The experimental data are as follows:

Test 1: Cutting speed = 300 ft/min (91.44 m/min), Tool life = 50 min

Test 2: Cutting speed = 400 ft/min (121.9 m/min), Tool life = 16 min

From the data given, solve for the constants in Eq. 3.4.

Substituting numerical values into Eq. 3.4 gives

$$300(50)^n = C \tag{1}$$

and

$$400(16)^n = C \tag{2}$$

Equating both equations enables the isolation of n as

$$300(50)^n = 400(16)^n$$
$$\left(\tfrac{50}{16}\right)^n = 1.333$$
$$n \ln 3.125 = \ln 1.333$$
$$n(1.1394) = 0.2874$$
$$n = 0.252$$

With the value of n, we can evaluate C by substituting into Eq. 1.

$$\begin{aligned} C &= 300(50)^{0.252} \\ &= 300(2.68) \\ &= 804 \end{aligned}$$

With the constants evaluated, the tool life equation for example 3.5 can be written as

$$V(T^{0.252}) = 804 \tag{3.5}$$

Another approach for determining the constants n and C of Eq. 3.4 is through a graphical layout. If the two experimental points of example 3.5 are plotted on log–log graph paper, then C (the y-intercept) and n (the slope of the curve) can be directly evaluated. Figure 3.8

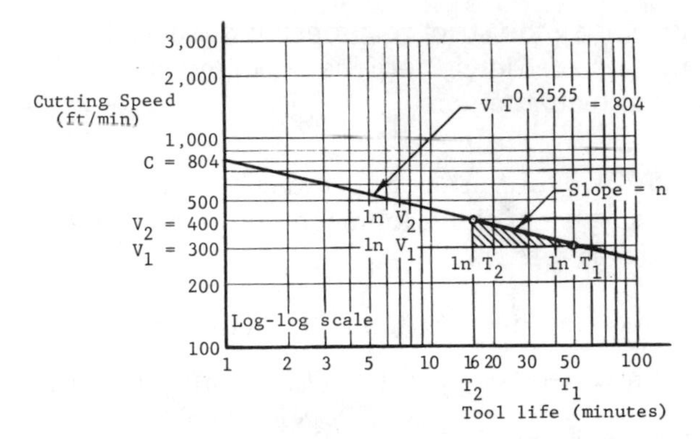

Fig. 3.8. Graphical representation of the tool life equation.

illustrates the graphical technique. From Fig. 3.8, n can be evaluated by finding the slope of the curve as follows:

$$n = \frac{\Delta \ln V}{\Delta \ln T}$$

$$= \frac{\ln V_2 - \ln V_1}{\ln T_1 - \ln T_2}$$

$$= \frac{\ln 400 - \ln 300}{\ln 50 - \ln 16}$$

$$= \frac{5.9915 - 5.7038}{3.9120 - 2.7726}$$

$$= 0.2525$$

A very close examination of Fig. 3.8 also reveals that the value of C (the y-intercept) can be evaluated as

$$C = 804$$

Once the analytical relationship between cutting speed and tool life is established as illustrated for the operation described by Eq. 3.5, cutting speeds for a specific tool life can be calculated in a range where the equation is valid. In other words, if there is confidence in the validity of the analytical expression, then it can be applied to calculate a recommended cutting speed that will yield a desired tool life. The following example illustrates such a case.

EXAMPLE 3.6. Given Eq. 3.5 as a description of the relationship between cutting speed and tool life, calculate the cutting speed that will yield a tool life of 30 min.

Table 3.4. Listing of Three Points from Tool-Life–Cutting-Speed Curve

Equation	Point	Cutting Speed (ft/min)	Cutting Speed (m/min)	Tool Life (min)
$V(T^{0.252}) = 804$	1	400	121.9	16
	2	341	103.9	30
	3	300	91.44	50

Substituting $T = 30$ into Eq. 3.5 gives

$$V = \frac{804}{(30)^{0.252}}$$

$$= 341 \text{ ft/min } (103.9 \text{ m/min})$$

Table 3.4 lists the three points from examples 3.5 and 3.6 and underscores the sensitive relationship between tool life and cutting speed. As can be seen from the data in Table 3.4, comparatively small changes in cutting speed have a profound effect on changing the tool life. To illustrate, reducing the cutting speed from 400 ft/min to 341 ft/min results in a 14.75% reduction in cutting speed; this changes the tool life from 16 min to 30 min, which is an 87.5% increase in tool life. The results are more dramatic between cutting speeds of 400 ft/min to 300 ft/min. A 25% reduction in cutting speed increases tool life from 16 min to 50 min, which is a 212.5% increase. The data of Table 3.4 are plotted in Fig. 3.9. Of special interest is a comparison between

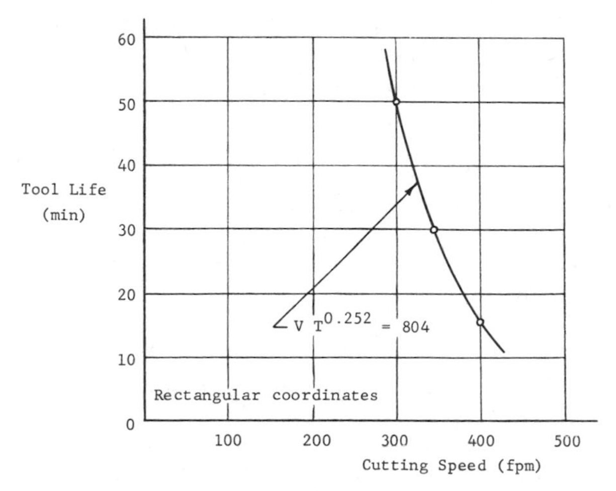

Fig. 3.9. Experimental tool-life–cutting-speed curve.

Fig. 3.9 and Fig. 3.5. The similarity of curves demonstrates the strong relationship between cutting temperature and cutting speed.

A final application of interest with regard to dealing with different tool materials is an evaluation of production results based on specific tool life equations. Example 3.7 lists the results of the comparison. It is noted that Taylor's equation with numerical values for n and C expresses the relationship between a tool material and a work material at a specific feed and depth of cut.

EXAMPLE 3.7. Two tool materials are described by the given tool life equations. Using these equations, calculate the corresponding cutting speeds for a recommended tool life of 45 min.

$$\text{Tool 1:} \qquad V(T^{0.25}) = 700$$

$$\text{Tool 2:} \qquad V(T^{0.17}) = 300$$

Substituting a tool life of 45 min into the given expressions yields

$$V_1 = \frac{700}{(45)^{0.25}} = 270 \text{ ft/min } (82.3 \text{ m/min})$$

$$V_2 = \frac{300}{(45)^{0.17}} = 157 \text{ ft/min } (47.9 \text{ m/min})$$

As can be seen from the results in example 3.7, tool 1 can be operated at a substantially higher cutting speed than tool 2 for the same tool life. As a result, if the machine setting is on the basis of the 45-min tool life, it can be expected that for the same feed and depth of cut, tool 1 will produce at a rate of 270/157 above the production of tool 2.

In the case where a comparison of the application of two tools is on the basis of performance at a given cutting speed, then tool life equations of the type given in example 3.7 can be used to calculate the corresponding tool life. Example 3.8 illustrates a case of this type.

EXAMPLE 3.8. Given the two tool life equations of example 3.7, determine the corresponding tool life for each tool for a cutting speed of 200 ft/min (60.96 m/min).

Substituting a cutting speed of 200 ft/min (60.96 m/min) into the given expressions yields

$$T_1 = \left(\tfrac{700}{200}\right)^{1/0.25} = 150.06 \text{ min}$$

$$T_2 = \left(\tfrac{300}{200}\right)^{1/0.17} = 10.86 \text{ min}$$

The difference in tool life for tool 1 and tool 2 in example 3.8 illustrates that although tool 2 can function at 200 ft/min (60.96 m/min), it requires a tool change every 10.86 min. The ratio of tool changes for this same operation between tool 1 and tool 2 is 150.06/10.86 or 13.82, a high comparative ratio.

In closing this section on examples of tool application, it can be stated that the best choice of a particular tool material for machining a given work material is not an easy one. Many factors need to be taken into consideration. These cannot always be defined in terms of simplified equations. Analytical expressions indicate that tool materials with high hot hardness display the best production capabilities. However, with increased hardness comes the vulnerability of brittleness, which leads to tool failure through chipping. In addition, other factors such as surface finish requirements may dictate the use of a tool material that performs in a desired fashion. In general, if tool life at a given cutting speed is a factor of consideration, positive results may be attained if a concentration is made on using tool materials that displays high hot hardness. Gains may also be attained by concentrating on techniques of reducing the cutting temperature.

3.11 SUMMARY OF CUTTING TOOL MATERIALS

The performance of a cutting tool is substantially dominated by the materials from which the tool is manufactured. The choice of materials from which the tool is to be made is dictated by the environment in which it is to function. In most cases, this involves conditions of elevated temperatures, high stresses, and strong abrasion. As a result, cutting tools must possess certain physical properties in order to perform successfully. These include: hardness characteristics for resisting high stress; toughness characteristics for absorbing impact without brittle fracture; thermal characteristics for retaining strength and hardness at elevated temperatures; and wear-resistant characteristics for preventing erosion of the cutting edge. If a tool can be kept efficiently operational, then it can be labeled as performing successfully. In order for this condition to exist, the tool material is usually matched to the work material in an application that produces good cutting conditions.

Tool materials can be classified into six general categories. These include carbon tool steels, high-speed steels, cast alloys, carbides, ceramics, and diamonds. The selection of a particular tool material depends on the work material to be cut as well as on the conditions (such as cutting speed) under which the cutting is to take place.

Carbon tool steels, made up basically of iron and carbon, can be successfully heat treated to control hardness, strength, and wear resistance. The great disadvantage of carbon tool steels lies with the characteristic of very poor hot hardness. As a result, these materials are limited in applications involving cutting tools used to a cut soft materials and in applications involving low speeds that do not generate elevated temperatures. Advantages of using carbon tool steels as tools center on the ability to retain a keen edge where high heat and high abrasion are not present. In addition, carbon tool steels display characteristics affiliated with easy machining and the retention of a tough core and a hard surface upon heat treatment.

High-speed steels are produced by adding specific alloying elements to iron to enable the material to retain hardness at elevated temperatures. These alloying elements include molybdenum, chromium, tungsten, vanadium, and cobalt. When compared to carbon tool steels, high-speed steels possess superior hot hardness as well as higher wear resistance. As a result, these tools function successfully under the adverse conditions of the cutting process, which includes elevated temperatures as well as high stress conditions.

Cast cobalt alloy tools, involving combinations of cobalt, chromium, tungsten, and carbon systems cast to shape, display high hardness and high toughness characteristics. Certain successful applications have been found in a cutting speed range that lies between high-speed tools and carbide tools. Cobalt tools usually perform well at higher cutting speeds than high-speed steels and possess more resistance to chipping than carbide tools.

Cemented carbide tools are characterized by their high hardness and ability to retain hardness at elevated temperatures. Consequently, they can be used successfully at higher speeds than high-speed steel tools. Another tool material that can be used successfully at higher cutting speeds is ceramic. Made up of aluminum oxides, oxides of other alloys, and refractory metal carbides, these materials exhibit high resistance to failure at elevated temperatures. As the hardness of tool materials increases, so does brittleness and the susceptibility to chipping. This limits the capability of the brittle materials in cases of interrupted cuts where high impact loads are present. As an example, diamond, the hardest known natural material, can be used successfully as a cutting tool material under proper conditions of very fine cuts on readily machinable work materials. However, diamond tools are not recommended for interrupted cutting that can cause chipping of the cutting edge. Figure 3.10 illustrates the general relationship between temperature and cutting speed for some major tool materials.

Surface treatment of tools can improve the cutting performance by increasing wear resistance through reduction of frictional forces that act between the tool and the work material. In addition, surface coat-

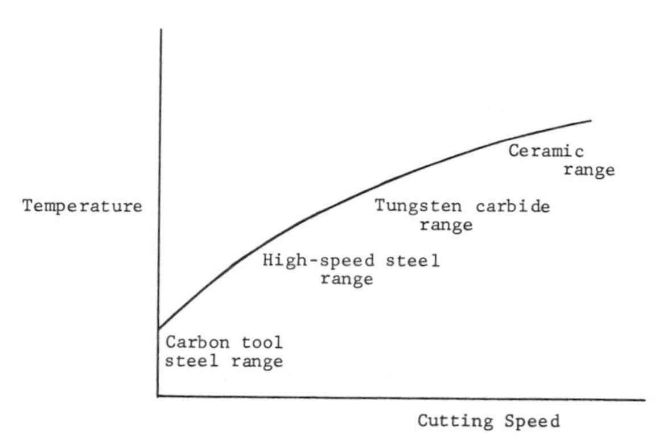

Fig. 3.10. Relationship between temperature and cutting speed for major tool materials.

ings can also serve as a barrier for heat flow from the chip to the tool. They can also serve as a means of reducing weldability of the work material with regard to the tool, thereby increasing the ability of the tool to perform successfully. Cutting fluids can also have a dramatic effect on improving tool performance. This is accomplished through the cooling effect of the cutting fluid on reducing the cutting temperature and by providing a lubricating effect on friction reduction between the tool and the work material.

A comparison of the performance of different tool materials on a given work material under specific cutting conditions can be revealing. Tool life, a measure of the time during which a tool functions satisfactorily, is found to be a function of the cutting temperature under which the tool functions. Therefore, tool materials with relatively low hot hardness have a tendency to also have a lower tool life. Since cutting temperature is difficult to measure, a more convenient factor in the tool life equation is the cutting speed, which is easy to measure. Taylor's tool life equation gives a description of the relationship of tool life with cutting speed. It serves as a means of comparing the performance of different tool materials in ranges where the equation is valid.

PROBLEMS

PROBLEM 3.1. Given three tool materials with the cutting-speed–tool-life relationship described as shown, determine the corresponding recommended cutting speeds for a tool life of 20 min.

Tool 1: $V(T^{0.5}) = 3{,}000$

Tool 2: $V(T^{0.25}) = 800$

Tool 3: $V(T^{0.125}) = 225$

Answers: $V_1 = 670$ ft/min
$V_2 = 378$ ft/min
$V_3 = 155$ ft/min

PROBLEM 3.2. For a metal cutting operation described by Eq. 3.3, calculate the tool life for an operating temperature of 950°F.

Answer: $T_L = 164.5$ min

PROBLEM 3.3. Determine the constants of Eq. 3.2 from the data compiled from the following two tests:

Test 1: $t_{p1} = 900°F$, $T_{L1} = 60$ min
Test 2: $t_{p2} = 1000°F$, $T_{L2} = 20$ min

Answers: $A = 4.089 \times 10^{32}$
$n = 10.437$

PROBLEM 3.4. Calculate the time required to machine a part with the following specifications. Turning operation: 76.2-mm diameter, 304.8-mm length with 1.5875-mm approach and 1.5875-mm overtravel allowance, feed of 0.3048 mm/rev, and cutting speed of 36.576 m/min.

Answer: $T_m = 6.61$ min

PROBLEM 3.5. Determine the change in the shear angle for a metal cutting operation as a result of applying a cutting fluid. The feed is 0.008 in./rev, depth of cut is 0.125 in., and the tool has an 8° rake angle. Without a cutting fluid, the chip thickness is 0.028 in., whereas with a cutting fluid the chip thickness measures 0.018 in.

Answer: $\Delta\alpha = 8.73°$
$(\alpha_1 = 16.4°;$
$\alpha_2 = 25.13°)$

PROBLEM 3.6. Two tests are conducted to determine the relationship between tool life and cutting speed for a given operation. In the first test, a cutting speed of 45.72 m/min results in a tool life of 60 min. In the second test, a cutting speed of 76.2 m/min results in a tool life of 20 min. Write Taylor's equation for this operation.

Answer: $V(T^{0.465}) = 306.63$,
where V is in units of m/min

PROBLEM 3.7. From the answer given in problem 3.6, calculate the cutting speed for a corresponding tool life of 30 min.

Answer: $V = 207$ ft/min

PROBLEM 3.8. A comparison of time required to turn a 76.3-mm-diameter steel shaft a length of 381 mm (including approach and over-travel) is to be made between two tools. The recommended cutting speed and feed for tool 1 are 33.53 m/min and 0.254 mm/rev. The recommended cutting speed and feed for tool 2 are 76.2 m/min and 0.254 mm/rev. Determine the difference in time required to machine the part using tool 2 as compared to tool 1.

Answer: $\Delta T = 5.99$ min
($T_1 = 10.7$ min;
$T_2 = 4.71$ min)

PROBLEM 3.9. A cutting operation is being examined under the conditions of running dry and running with a cutting fluid with a tool that has a 15° rake angle. The feed is set at 0.015 in./rev with a depth of cut of 0.125 in. The dry operation produces a chip thickness of 0.045 in., whereas the cutting fluid operation produces a chip thickness of 0.030 in. Determine the difference in shear area between the dry and the wet operation.

Answer: $\Delta A_s = 0.001783$ in.^{2}
($A_{sd} = 0.005648$ in.^{2};
$A_{sw} = 0.003865$ in.^{2})

PROBLEM 3.10. From the guide given in problem 3.9, determine the shear force for the dry and wet applications for a work material that has a 172.4×10^6-Pa shear yield stress.

Answers: $\mathbf{F}_d = 628$ N
$\mathbf{F}_w = 430$ N

PROBLEM 3.11. If the relationship between cutting speed and tool life for a high-speed steel tool machining cast iron is $V(T^{0.111}) = 185$, determine the recommended cutting speed for a tool life of 45 min.

Answer: $V = 121$ ft/min

PROBLEM 3.12. Figure 2.11 illustrates the resolution of the force that the tool exerts on the work material into two perpendicular components, a normal and a frictional force. For a cutting application with an uncoated tool, the resultant cutting force is 1557 N. This is the vector sum of a normal force of 1112 N and a frictional force of 1090 N. The use of a coated tool has the effect of reducing the frictional force by

25%. It is reduced from 1090 N to 817 N. Determine the resulting cutting force using the coated tool, under the assumption that the normal force remains unchanged at 1112 N.

Answer: $\mathbf{F}_R = 1380$ N

PROBLEM 3.13. Two tool materials are described by the given tool life equations. Using these equations, calculate the corresponding cutting speeds for a tool life of 30 min.

Tool 1: $V(T^{0.30}) = 650$ Note: V is given in units of ft/min
Tool 2: $V(T^{0.20}) = 400$

Answers: $V_1 = 234$ ft/min
$V_2 = 202$ ft/min

PROBLEM 3.14. Given the two tool life equations of problem 3.13, determine the corresponding tool life for each tool for a cutting speed of 76.2 m/min. Note: The tool life equations must be rewritten in units of m/min.

Answers: $T_1 = 24.17$ min
$T_2 = 10.45$ min

PROBLEM 3.15. Write an equation correlating the cutting temperature and the cutting speed by isolating the tool life in both Eq. 3.2 and Eq. 3.4 and equating both terms. Note that the exponents n in Eq. 3.2 and in Eq. 3.4 are different values.

Answer: $$t_p = (A)^{1/n_1}\left(\frac{V}{C}\right)^{1/n_1 n_2}$$

from $T = \dfrac{A}{(t_p)^{n_1}}$; $(T)^{n_2} = \dfrac{C}{V}$

BIBLIOGRAPHY

Bhattacharyya, Amitabha, and Ham, Inyong, *Design of Cutting Tools*, American Society of Tool and Manufacturing Engineers, Dearborn, Michigan, 1969.

Christopher, John D., and Zlatin, Norman, *New Cutting Tool Materials*, Society of Manufacturing Engineers, Dearborn, Michigan, 1974.

Dallas, Daniel B., *Tool and Manufacturing Engineers Handbook*, Society of Manufacturing Engineers, Dearborn, Michigan, 1976.

DeVries, Marvin F., *Machining Economics—A Review of the Traditional Approaches and Introduction to New Concepts*, American Society of Tool and Manufacturing Engineers, Dearborn, Michigan, 1969.

Ham, I, and Narutaki, N., *Wear Characteristics of Ceramic Tools*, American Society of Mechanical Engineers, New York, 1972.

Holtz, Frederick C., *New Developments in Cutting Tool Materials*, American Society of Tool and Manufacturing Engineers, Dearborn, Michigan, 1969.

Howe, Raymond E., *Producibility/Machinability of Space-Age and Conventional Materials*, American Society of Tool and Manufacturing Engineers, Dearborn, Michigan, 1968.

Johnson, Carl G., *Metallurgy*, American Technical Society, Chicago, Illinois, 1956.

Kalish, Herbert S., "An Update in New Developments in Cutting Tools," *Manuf. Eng.*, September 1978.

Machining Data Handbook, Metcut Research Associates, Cincinnati, Ohio, 1972.

Manual on Cutting Metal, American Society of Mechanical Engineers, New York, 1952.

Sen, Gopal Chandra, and Bhattacharyya, Amitabha, *Principles of Metal Cutting*, New Central Book Agency, Calcutta, India, 1969.

Shaw, Milton C., *Metal Cutting Principles*, M.I.T. Press, Cambridge, Massachusetts, 1968.

Swinehart, Haldon J., *Cutting Tool Material Selection*, American Society of Tool and Manufacturing Engineers, Dearborn, Michigan, 1968.

Taylor, F. W., *On the Art of Cutting Metals*, American Society of Mechanical Engineers, New York, 1906.

Whitney, E. Dow, *Ceramic Tools*, Society of Manufacturing Engineers, Dearborn, Michigan, 1973.

4

Mechanics of the Cutting Process

4.1 INTRODUCTION

The mechanics of the cutting process deals primarily with the analysis of the forces acting between the tool and the work material. By analyzing forces that act in particular directions relative to the tool and work material, we can gain insight into a basic understanding of the cutting process. In this chapter, the parallelogram law for the addition and resolution of vectors serves as a means of convenient placement of component forces for purposes of analysis of specific directional factors such as friction and shear. This law states that a resultant force can be resolved into two component forces such that the resultant force is equal to the diagonal of a parallelogram and that the component forces are equal to the sides of the parallelogram. The law graphically depicts the vectorial qualities of forces that must take into account not only the magnitude but also the direction of the force. Figure 2.19 illustrates the parallelogram law.

Newton's first and third general laws of mechanics also play an important role in the force analysis of the cutting process. The first law is that of equilibrium. It states that if a body is at rest or is moving in a straight line at a constant velocity, then a vectorial summation of all the forces acting on the body is equal to zero. In other words, the force system (summation of the forces) is balanced in a state of equilibrium where the resultant of all the forces acting on the body is equal to zero.

Newton's third law is the law of mechanics dealing with action and reaction. It states that when two bodies exert forces on each other, these forces are equal in magnitude, opposite in direction, and act in the same line of action. The application of the laws of mechanics is of prime importance in the analysis of the cutting process. This is done through the drawing of free-body diagrams from which analytical relationships can be written. A free-body diagram is a drawing of a body showing in vector form all the forces acting on the body. Figure 4.1 illustrates a free-body diagram of a tool showing the force that a workpiece exerts on the tool in balance with the reaction force that the tool-holder exerts on the tool. The tool is in equilibrium, that is, the summation of the action (workpiece force) is balanced with the reaction (tool-holder force).

Of importance in the mechanics of the cutting process is a description of the force interaction of the tool with the work material. It is through an analysis of the forces acting during the metal cutting process that this description can be made. Cutting forces reflect the effectiveness of the cutting tool as well as the resistance that the work material offers to the cutting operation. Analysis can be enhanced through the resolution of the resultant cutting force into convenient components. As an example, the component along the face of the tool is the *frictional force*, and through it the surface characteristic known as the *coefficient of friction* can be evaluated. The component of the resultant force acting in the direction of the shear plane can be used to evaluate the stress acting over the shear plane.

Isolation of a component force in a particular direction places attention on a specific segment of the cutting operation. To cite an example, a measure of the horsepower required in a cutting operation can conveniently be evaluated by isolating the component of the resultant force in the direction of the motion and multiplying it by the cutting speed.

Topics covered in this chapter take into account the application of the components of the resultant force acting on the tool and the chip. Tool forces are first considered in terms of the component in the direction of the cutting speed and the component in the direction of the feed. This is followed by an examination of the component forces along the

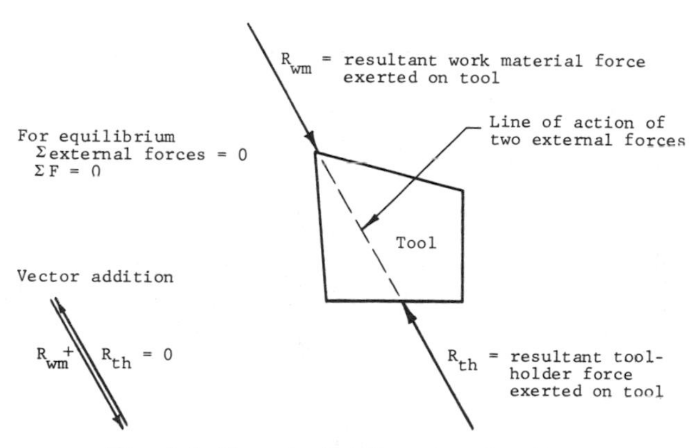

Fig. 4.1. Free-body diagram of tool.

tool–chip interface as well as an examination of the component forces acting on the chip–work material interface, that is, on the shear plane. A combination of these component forces then leads to the analysis of three sets of perpendicular force components that are all geometrically related to the resultant cutting force.

Another topic introduced in this chapter is the coefficient of friction between the tool and the chip. The results of an experiment designed to measure the coefficient of friction as a function of cutting speed are also listed. This is followed by an analysis of the velocity relationships among the tool, chip, and work material. With the availability of velocity and force relationships, energy considerations are then made that lead to an analysis of power consumption in the cutting process.

Some of the latter topics covered in this chapter deal with the effective tool clearance angles due to tool positioning as well as those due to the relative velocity of the tool to the workpiece. The final topic is a three-dimensional analysis of the cutting force, accompanied by the associated analytical expressions.

4.2 TOOL FORCE

When cutting force measurements are taken, usually force-sensing instruments (dynamometers) are used to measure directional components of the resultant force acting on the tool. Figure 4.2 illustrates a dynamometer designed to measure a vertical and a horizontal force.

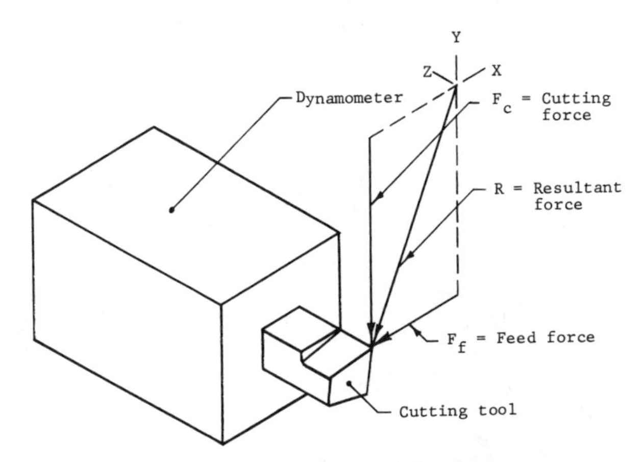

Fig. 4.2. Dynamometer with cutting and feed forces acting on tool.

In a turning operation, the vertical force is in the direction of the cutting speed and is referred to as the *cutting force* ($\mathbf{F}_c$). The horizontal force is in the direction of the feed and is referred to as the *feed force* ($\mathbf{F}_f$). The vector sum of the cutting force and the feed force is equal to the resultant force that the tool exerts on the work material. Figure 4.3 shows the resultant force resolved into two components. The action–reaction principle of Newton's third law of mechanics is illustrated in the diagram. If the force (**R**) that the tool exerts on the work material is considered the action, then the reaction to this force which the work material exerts on the tool is equal to the force vector labeled **R**′ in Fig. 4.3.

Of interest is the determination of the magnitude and direction of the resultant force acting between the tool and work material. If dynamometer readings are taken for a given operation that reveal the magnitude of the cutting force $\mathbf{F}_c$ and the feed force $\mathbf{F}_f$, then trigonometric relationships can be formed to determine the resultant force. From Fig. 4.3 it can be written that

$$\mathbf{F}_c = \mathbf{R}\cos\theta \tag{4.1}$$

$$\mathbf{F}_f = \mathbf{R}\sin\theta \tag{4.2}$$

$$|\mathbf{R}| = \sqrt{(\mathbf{F}_c)^2 + (\mathbf{F}_f)^2} \tag{4.3}$$

and

$$\tan\theta = \frac{\mathbf{F}_f}{\mathbf{F}_c} \tag{4.4}$$

The following example illustrates how the magnitude and direction of the resultant cutting force can be determined from dynamomter readings of the cutting force and the feed force.

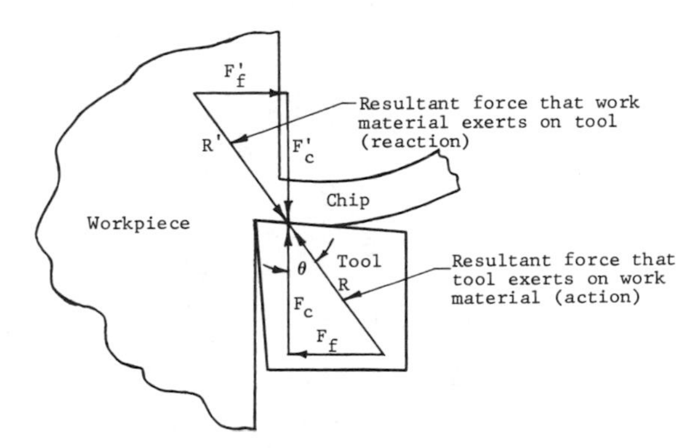

Fig. 4.3. Resultant forces acting in metal cutting process.

EXAMPLE 4.1. Given dynamometer readings where $\mathbf{F}_c = 250$ lb (1112 N) and $\mathbf{F}_f = 75$ lb (333.6 N) calculate the magnitude and direction of the resultant force that the tool exerts on the work material.

Substituting values into Eq. 4.3 yields

$$|\mathbf{R}| = \sqrt{(250)^2 + (75)^2} = 261 \text{ lb} (1161 \text{ N})$$

Substituting values into Eq. 4.4 yields

$$\theta = \tan^{-1}\frac{75}{250} = 16.7°$$

EXAMPLE 4.2. In a cutting operation similar to that shown in Fig. 4.3, the force that the work material exerts on the tool ($\mathbf{R}'$) has a magnitude of 200 lb (890 N) and a 20° angular displacement (θ) from the vertical direction. From these data, determine the magnitude of the cutting force ($\mathbf{F}_c$) as well as the magnitude of the feed force ($\mathbf{F}_f$).

Substituting the given values into Eqs. 4.1 and 4.2 yields

$$\mathbf{F}_c = 200 \cos 20° = 187.9 \text{ lb} (835.8 \text{ N})$$

$$\mathbf{F}_f = 200 \sin 20° = 68.4 \text{ lb} (304.3 \text{ N})$$

4.3 TOOL–CHIP INTERFACE FORCE

The force acting between the tool and the chip during the metal cutting operation is equal to the resultant force that the tool exerts on the work material. This resultant force can be resolved into two convenient and useful components. Figure 4.4 illustrates the resolution of

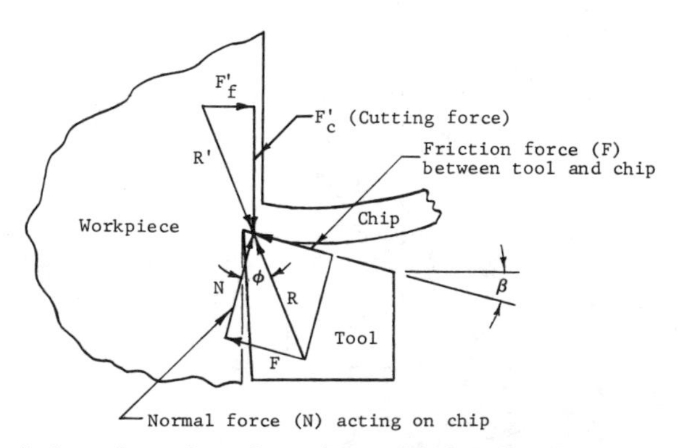

Fig. 4.4. Resolution of resultant force into a frictional and a normal component.

the resultant force into a component parallel to the face of the tool and a second component that is perpendicular to the face of the tool. The component parallel to the face of the tool gives a measure of the resistance that the surface provides to the sliding motion of the chip. It is referred to as the *frictional force* (**F**). The component perpendicular to the face of the tool is referred to as the normal force (**N**) and it provides a measure of the compressive action of the tool on the chip.

When two surfaces are in contact, as in the case of the tool with the chip (Fig. 4.4), and one surface is made to move with respect to the other surface, forces tangent to the surface develop. These forces are frictional forces and are a reflection of the quality of the surface with respect to sliding action. A frictional force is the resistance that two bodies provide with regard to sliding motion. An experiment on sliding friction reveals that there is a relationship between the normal force and the frictional force. In engineering mechanics, this relationship is called the *coefficient of friction.* It is expressed as

$$\mu = \frac{\mathbf{F}}{\mathbf{N}} = \frac{\text{Frictional force}}{\text{Normal force}} \tag{4.5}$$

An examination of Fig. 4.4 reveals that this ratio can be expressed in terms of the angle ϕ as

$$\mu = \tan\phi = \frac{\mathbf{F}}{\mathbf{N}} \tag{4.6}$$

For sliding motion, the angle ϕ is defined as the angle of kinetic friction.

As can be seen from Eq. 4.5, the coefficient of friction is defined as the ratio of the frictional force to the force acting normal to the surface. In metal cutting, this ratio is affected by the quality of the tool surface as well as by the interaction of the chip and the tool. The velocity of the chip over the tool surface has an influence on the magnitude of the coefficient of friction as does the temperature of the chip. When a comparison is made with low-speed ordinary sliding, the coefficient of friction associated with chip sliding becomes a relatively complicated term. This can be attributed to factors such as the mechanical interlocking of the surface irregularities (asperites) due to the high normal forces as well as to the tendency under conditions of elevated temperatures and high normal forces of the chip to weld itself to the tool surface. The latter condition is the cause of the built-up edge on the tool, which demands a shearing action of the metal chip as it flows over the tool. A condition of this type can cause a large increase in the value of the coefficient of friction.

Figure 4.5 is an illustration of the action–reaction relationship of the frictional and normal forces acting between the chip and the tool.

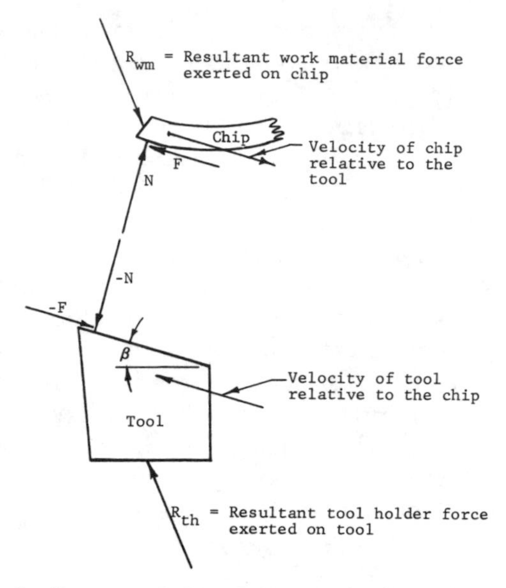

Fig. 4.5. Free-body diagram of chip and tool illustrating normal and frictional forces at the interface.

In the diagram, the chip is separated from the tool to more clearly describe the direction of the forces. Note that the frictional forces **F** and −**F** act in a direction opposite to the direction of sliding, that is, the chip is sliding over the tool in a direction opposite to the direction of the force **F**, whereas the tool is sliding with a motion relative to the chip in a direction opposite to the direction of the force −**F**. In other words, if one took a ride on the chip, it would appear, from the chip, that the tool is moving in the direction of the force **F**. From this, the conclusion can be drawn that the frictional force **F** is the force that the tool exerts on the chip in resistance to the sliding of the chip over the tool. In turn, the frictional force −**F** is the reaction to the force **F**. It is equal in magnitude, but opposite in direction, to the force **F**. The frictional force −**F** is the force that the chip exerts on the tool in an effort to resist the relative motion (sliding) of the tool to the chip.

The frictional and normal forces shown in Fig. 4.5 are internal forces that act on the tool–chip interface. They appear in Fig. 4.1 in the form of the resultant forces acting on the tool through the work material as well as through the tool holder. A vector addition of these internal forces representing the action–reaction forces acting between the tool and the chip is shown in Fig. 4.6.

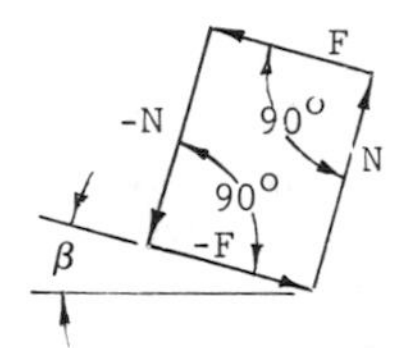

Fig. 4.6. Vector equilibrium of action–reaction forces acting between tool and chip.

This vector summation can be expressed as

$$\mathbf{N} \nrightarrow \mathbf{F} \nrightarrow (-\mathbf{N}) \nrightarrow (-\mathbf{F}) = 0 \tag{4.7}$$

yielding an equilibrium condition where the sum of the vectors is equal to zero.

If the forces acting on the tool are examined, it becomes apparent, as shown in Fig. 4.5, that the summation can be written as

$$\mathbf{R}_{\mathbf{th}} \nrightarrow (-\mathbf{N}) \nrightarrow (-\mathbf{F}) = 0 \tag{4.8}$$

In a similar fashion, the chip forces can be added in the form

$$\mathbf{R}_{\mathbf{wm}} \nrightarrow \mathbf{N} \nrightarrow \mathbf{F} = 0 \tag{4.9}$$

Figure 4.7 illustrates a graphical representation of Eqs. 4.8 and 4.9. Additional expressions that can be written are

$$\mathbf{N} = \mathbf{R} \cos \phi \tag{4.10}$$

$$\mathbf{F} = \mathbf{R} \sin \phi \tag{4.11}$$

and

$$|\mathbf{R}| = \sqrt{\mathbf{F}^2 + \mathbf{N}^2} \tag{4.12}$$

The following numerical examples demonstrate some trigonometric manipulation dealing with the interface forces between the tool and the chip.

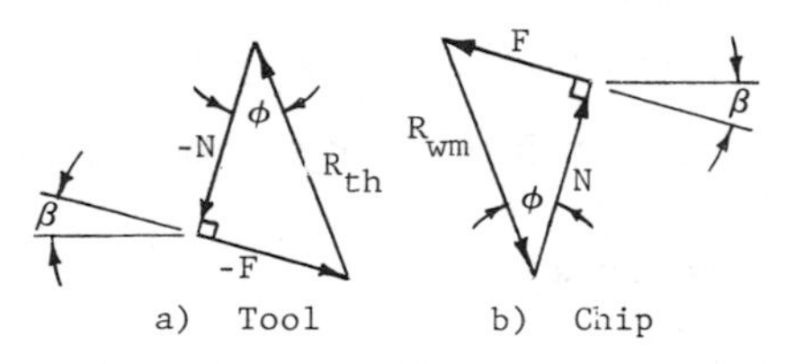

Fig. 4.7. Vector summation of external forces acting on (a) chip and (b) tool.

EXAMPLE 4.3. Given a resultant force of 261 lb (1161 N) and an angular orientation of 16.7°, as illustrated in example 4.1, determine the frictional and normal forces acting between the tool and the chip for the case where the rake angle on the tool is 10°. In addition, determine the coefficient of friction for this operation.

The following sketch can be drawn from Fig. 4.7(b):

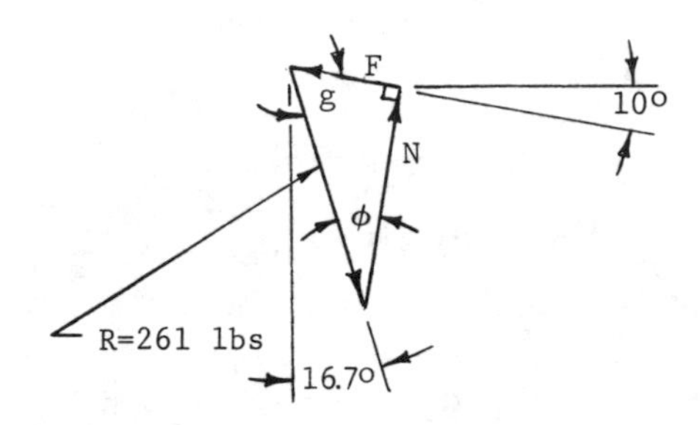

Angle g can be determined by noting that it is part of a right angle and can be expressed as

$$90° = 16.7° + g + 10°$$
$$g = 63.3°$$

Noting that the vector polygon is a right triangle, where the sum of the interior angles is equal to 180°, the angle of kinetic friction can be evaluated as

$$180° = 90° + g + \phi$$
$$\phi = 26.7°$$

Using the angle of kinetic friction, the normal force can be written as

$$\mathbf{N} = \mathbf{R}\cos\phi$$
$$= 261(0.893) = 233.2 \text{ lb} (1037 \text{ N})$$

In a similar fashion, the friction force can be written as

$$\mathbf{F} = \mathbf{R}\sin\phi$$
$$= 261(0.4493) = 117.27 \text{ lb} (521.6 \text{ N})$$

Solving for the coefficient of friction, it can be written that

$$\mu = \tan\phi = \frac{\mathbf{F}}{\mathbf{N}}$$

$$= \tan 26.7° = \frac{117.27}{233.2} = 0.503$$

EXAMPLE 4.4. For a given metal cutting operation, dynamometer readings reveal values of 300 lb (1334 N) for the cutting force $\mathbf{F}_c$ and 100 lb (444.8 N) for the feed force $\mathbf{F}_f$. The cutting tool has a positive rake angle of 12°. From these data, calculate the value of the frictional force, the normal force, and the coefficient of friction.

A diagram of the tool forces involved in the calculation is of the form

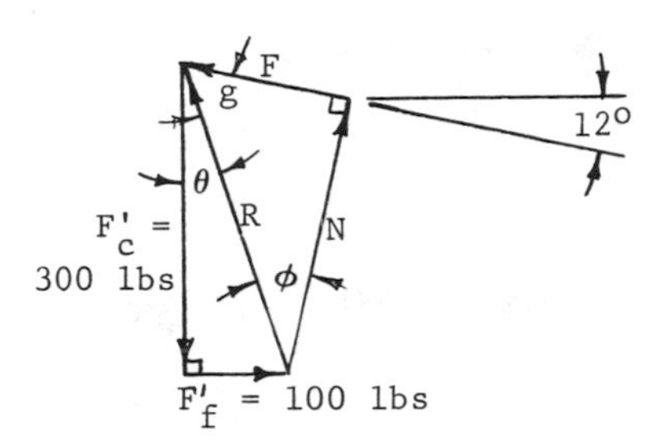

Solving for θ and $\mathbf{R}$ yields

$$\tan\theta = \frac{\mathbf{F}'_f}{\mathbf{F}'_c} = \frac{100}{300} = 0.333$$

$$\theta = 18.4^\circ$$

from which

$$\mathbf{R} = \frac{\mathbf{F}'_c}{\cos\theta} = \frac{300}{0.9487} = 316.2 \text{ lb} (1406.5 \text{ N})$$

Noting that a common hypotenuse is shared by the two right triangles listed above, it can be concluded that the solution to this problem lies in evaluating angles g and ϕ:

$$\begin{aligned} g &= 90^\circ - \theta - \beta = 90^\circ - 18.4^\circ - 12^\circ \\ &= 59.6^\circ \end{aligned}$$

and

$$\begin{aligned} \phi &= 90^\circ - g = 90^\circ - 59.6^\circ \\ &= 30.4^\circ \end{aligned}$$

The coefficient of friction can now be written as

$$\mu = \tan\phi = \frac{\mathbf{F}}{\mathbf{N}}$$

$$= 0.587$$

The frictional force can be evaluated as

$$\begin{aligned}\mathbf{F} &= \mathbf{R}\sin\phi \\ &= 316.2\sin 30.4^\circ \\ &= 160\ \text{lb}\ (711.7\ \text{N})\end{aligned}$$

and the normal force can be written as

$$\begin{aligned}\mathbf{N} &= \mathbf{R}\cos\phi \\ &= 316.2\cos 30.4^\circ \\ &= 272.7\ \text{lb}\ (1213\ \text{N})\end{aligned}$$

4.4 SHEAR PLANE FORCES

The resultant force that the work material exerts on the tool during the metal cutting operation is transmitted through the shear zone between the tool and the chip. For purposes of analysis, this resultant force can be resolved into two convenient components that act on the shear zone. As shown in Fig. 4.8, the component parallel to the shear zone can be considered as the force acting in the direction of the shear. It is called the *shear force* ($\mathbf{F}_s$). The second component, acting normal to the shear plane, provides a measure of the compressive force acting on the shear plane. It is called the *normal force* ($\mathbf{F}_n$) acting on the shear plane. In Fig. 4.8, the chip is separated from the work material along the shear zone in order to demonstrate the internal shear and normal forces. A vector addition of the chip–work-material interface forces is shown in Fig. 4.9. This diagram verifies the equilibrium condition of the internal forces, where the summation of the vectors is

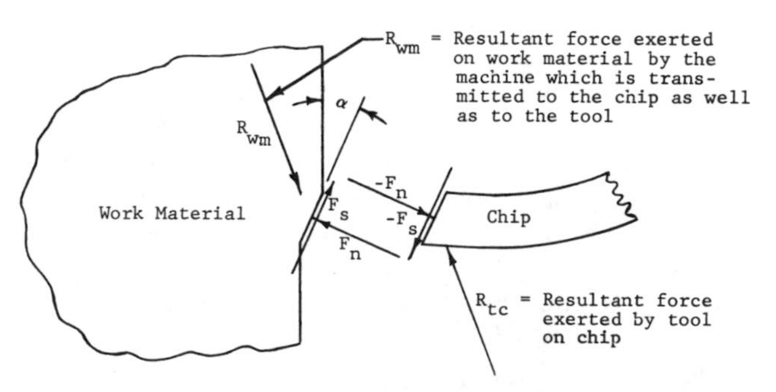

Fig. 4.8. Free-body diagrams of work material and chip, illustrating shear force and force acting normal to shear plane.

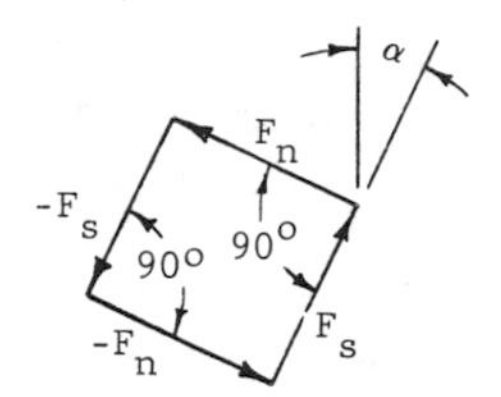

Fig. 4.9. Vector equilibrium diagram of action–reaction forces acting between the chip and the work material at the shear plane.

equal to zero. A vector summation of Fig. 4.9 can be expressed as

$$\mathbf{F}_s \mathrel{+\!\!\!\rightarrow} \mathbf{F}_n \mathrel{+\!\!\!\rightarrow} -\mathbf{F}_s \mathrel{+\!\!\!\rightarrow} -\mathbf{F}_n = 0 \tag{4.13}$$

In Fig. 4.8, the work material and the chip are shown as free bodies. For the sake of analysis, the forces acting on each body can be considered as external, and an equilibrium force balance can be expressed in vector form. This is illustrated in Fig. 4.10.

A vector summation of the graphical display of Fig. 4.10 can be expressed as

$$-\mathbf{F}_s \mathrel{+\!\!\!\rightarrow} -\mathbf{F}_n \mathrel{+\!\!\!\rightarrow} \mathbf{R}_{tc} = 0 \tag{4.14}$$

and

$$\mathbf{F}_s \mathrel{+\!\!\!\rightarrow} \mathbf{F}_n \mathrel{+\!\!\!\rightarrow} \mathbf{R}_{wm} = 0 \tag{4.15}$$

In terms of the resultant force, the shear force and the force normal to the shear plane can be written as

$$\mathbf{F}_s = \mathbf{R} \cos h \tag{4.16}$$

$$\mathbf{F}_n = \mathbf{R} \sin h \tag{4.17}$$

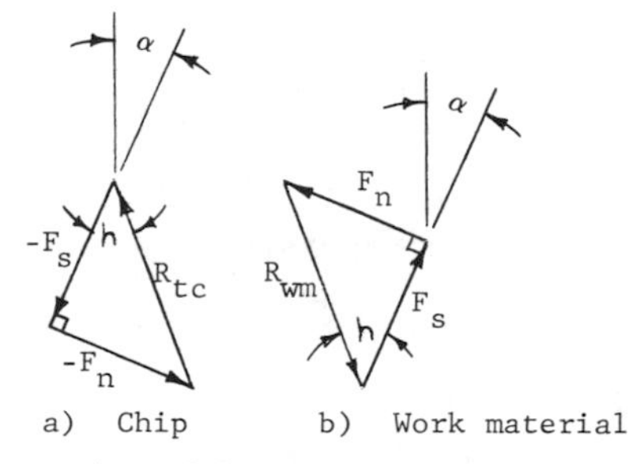

Fig. 4.10. Vector summation of the equilibrium external forces acting on (a) the chip and (b) the work material.

The magnitude of the resultant force can be expressed in terms of the shear force and the force normal to the shear plane as

$$|\mathbf{R}| = \sqrt{(\mathbf{F}_s)^2 + (\mathbf{F}_n)^2} \tag{4.18}$$

EXAMPLE 4.5. Given the resultant force of 261 lb (1161 N) with an angular orientation of 16.7°, as illustrated in examples 4.1 and 4.3, determine the shear force and the force normal to the shear plane for the metal cutting operation. To measure the shear angle α, the chip thickness was measured. It was found to be 0.045 in. (1.143 mm) for a corresponding feed of 0.015 in./rev (0.381 mm). The tool rake angle is 10°.

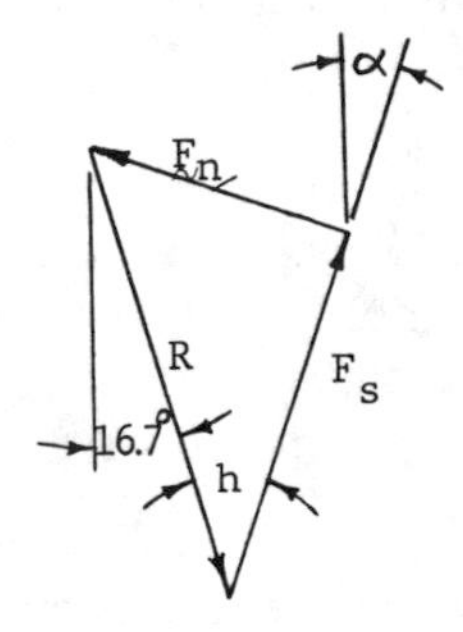

Solving for the shear angle from Eq. 2.22 yields

$$\tan\alpha = \frac{r_a \cos\beta}{1 - r_a \sin\beta} = \frac{0.33\cos 10°}{1 - 0.33\sin 10°}$$

$$= 0.3448$$

$$\alpha = 19.02°$$

From the diagram shown above, the angle h can be evaluated as

$$h = 16.7° + \alpha$$
$$= 35.72°$$

The shear force can now be written as

$$\mathbf{F}_s = \mathbf{R}\cos h$$
$$= 261 \cos 35.72°$$
$$= 211.9 \text{ lb } (942.6 \text{ N})$$

The force normal to the shear plane is then equal to

$$\mathbf{F}_n = \mathbf{R}\sin h$$
$$= 261 \sin 35.72°$$
$$= 152.4 \text{ lb } (677.9 \text{ N})$$

EXAMPLE 4.6. For the same metal cutting operation as described in example 4.4, it is observed that the shear plane angle α is equal to 15°. With dynamometer readings of 300 lb (1334 N) for the cutting force $\mathbf{F}_c$ and 100 lb (444.8 N) for the feed force $\mathbf{F}_f$, determine the shear force and the force normal to the shear plane.

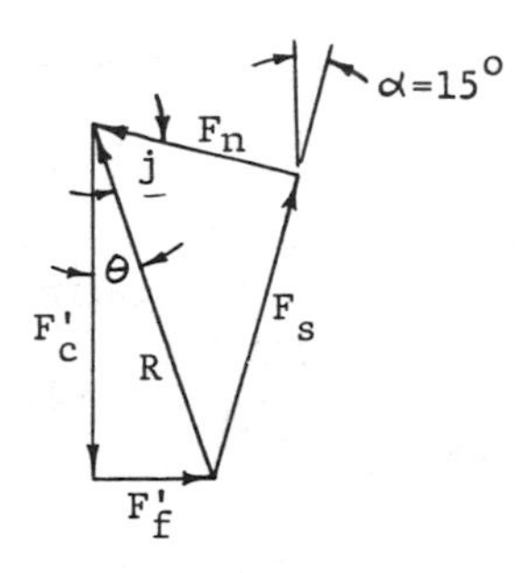

From the diagram listed above, it can be written that

$$j = 90° - \theta - \alpha$$

where $\theta = 18.4°$ from example 4.4. As a result,

$$j = 56.6°$$

In terms of the resultant force that was calculated in example 4.4, it can be written that

$$\begin{aligned}\mathbf{F}_s &= \mathbf{R} \sin j \\ &= 316.2 \sin 56.6° \\ &= 264 \text{ lb } (1174 \text{ N})\end{aligned}$$

and that

$$\begin{aligned}\mathbf{F}_n &= \mathbf{R} \cos j \\ &= 316.2 \cos 56.6° \\ &= 174 \text{ lb } (774 \text{ N})\end{aligned}$$

A point of interest is that example 4.6 can be solved directly from the given cutting and feed forces by adding vector components in the x and y directions. As an example, from the diagram listed above it can be written that

$$\mathbf{F}_s \mapsto \mathbf{F}_n \mapsto \mathbf{F}'_c \mapsto \mathbf{F}'_f = 0 \tag{4.19}$$

These four forces are in equilibrium, that is, their vector summation is equal to zero. The conclusion can be reached that the summation of the components of these vectors in any direction must also be equal to zero. Therefore,

$$F_{sx} \nleftrightarrow F_{nx} \nleftrightarrow F'_{cx} \nleftrightarrow F'_{fx} = 0 \tag{4.20}$$

and

$$F_{sy} \nleftrightarrow F_{ny} \nleftrightarrow F'_{cy} \nleftrightarrow F'_{fy} = 0 \tag{4.21}$$

Figure 4.11 illustrates the resolution of vectors into convenient components in the x and y directions.

An illustrated example of the application of the x and y components in solving a force problem is listed below as a confirmation of the results of example 4.6. Using force components as shown in Fig. 4.11(b), the shear force and the force normal to the shear plane can easily be evaluated. It is noted from example 4.6 that $\mathbf{F}'_c = 300$ lb, $\mathbf{F}'_f = 100$ lb, and $\alpha = 15°$. This is symbolically illustrated in Fig. 4.11(a). Taking components in the x direction yields

$$\mathbf{F}'_f + F_{sx} - F_{nx} = 0 \tag{4.22}$$

Substituting gives

$$100 + \mathbf{F}_s \sin 15° - \mathbf{F}_n \cos 15° = 0 \tag{4.22a}$$

In a similar fashion, taking components in the y direction yields

$$-\mathbf{F}'_c + F_{sy} + F_{ny} = 0 \tag{4.23}$$

Substituting gives

$$-300 + \mathbf{F}_s \cos 15° + \mathbf{F}_n \sin 15° = 0 \tag{4.23a}$$

As can be seen, the result of the substitution is a set of two simultaneous

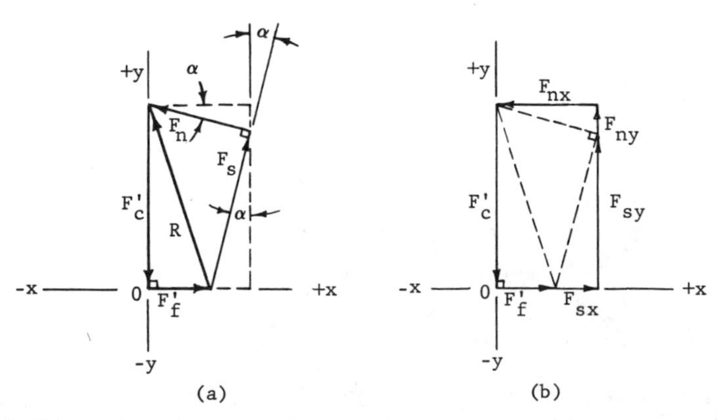

Fig. 4.11. Resolution of vectors into convenient components along the *x* and *y* axes.

equations containing the two unknown values of $\mathbf{F}_s$ and $\mathbf{F}_n$. An evaluation can be performed in the following manner:

Rewriting Eqs. 4.22a and 4.23a gives

$$100 + 0.2588\mathbf{F}_s - 0.9659\mathbf{F}_n = 0 \quad \text{(4.22b)}$$

$$-300 + 0.9659\mathbf{F}_s + 0.2588\mathbf{F}_n = 0 \quad \text{(4.23b)}$$

Multiplying Eq. 4.22b by the ratio 0.2588/0.9659 and adding the equations eliminates the $\mathbf{F}_n$ term and yields

$$-273.21 + 1.0352\mathbf{F}_s = 0$$

from which

$$\mathbf{F}_s = 264 \text{ lb } (1174 \text{ N})$$

Substituting 264 lb for $\mathbf{F}_s$ in either Eq. 4.22b or Eq. 4.23b enables the evaluation of $\mathbf{F}_n$ as

$$\mathbf{F}_n = 174 \text{ lb } (774 \text{ N})$$

4.5 THREE SETS OF PERPENDICULAR FORCE COMPONENTS

The resultant force that the tool exerts on the work material can be conveniently resolved into three sets of useful perpendicular components. When analyzed individually, each of these components represents the projection of the resultant force in a particular direction. For example, a projection of the resultant force onto a direction parallel to the face of the tool gives a measure of the frictional force (**F**) between the tool and the chip. The projection of the resultant force onto the shear plane gives a measure of the force acting in the direction of the shearing action ($\mathbf{F}_s$) between the chip and the work material. In a similar fashion, the cutting force ($\mathbf{F}_c$) can be determined by projecting the resultant force onto a plane parallel to the direction of the cutting velocity.

Figure 4.12 illustrates the resultant force (**R**) that the tool exerts on the work material. The resultant force (**R**) is shown in Fig. 4.12 as the diameter of a circle. If straight lines are extended from the ends of the diameter (from points *O* and *B*) in such a way as to intersect on the circle (point *A*), then a right triangle will be formed. The diameter serves as the hypotenuse of the triangle and in every case the angle between the lines drawn from points *O* and *B* will form a 90° angle.

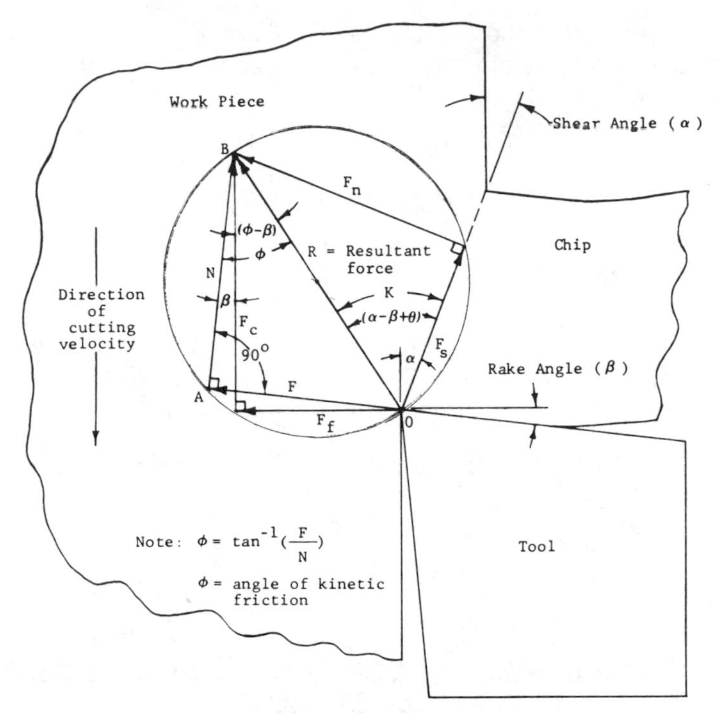

Fig. 4.12. Resolution of resultant force into three sets of perpendicular components.

An examination of Fig. 4.12 reveals that the resultant force serving as the diameter of the circle also serves as the common hypotenuse of three sets of perpendicular components. By identifying an interior angle of each of the three right triangles, one can evaluate the individual components as a function of the resultant force (**R**). The three interior angles are given in Fig. 4.12 in terms of the angle of kinetic friction (ϕ), the shear angle (α), and the rake angle (β). In order to determine the resultant force, usually dynamometer readings in terms of the cutting force ($\mathbf{F}_c$) and the feed force ($\mathbf{F}_f$) are taken. Figure 4.13 illustrates a tool with the resultant force resolved into two components, the cutting force ($\mathbf{F}_c$) and the feed force ($\mathbf{F}_f$).

The magnitude of the resultant force can be written as

$$|\mathbf{R}| = \sqrt{(\mathbf{F}_c)^2 + (\mathbf{F}_f)^2} \tag{4.24}$$

Once the resultant force is determined, trigonometric relationships enable us to determine the other force components. Since the rake angle (β) is a known quantity for a particular cutting operation, then

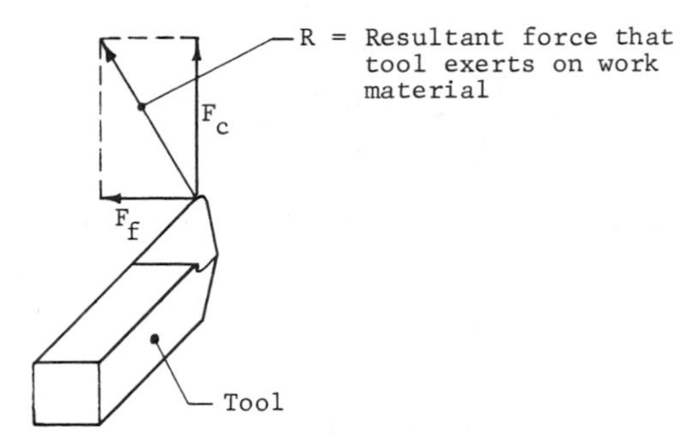

Fig. 4.13. Resultant force resolved into two components.

the angle $\phi - \beta$ shown in Fig. 4.12 can be determined by writing

$$\tan(\phi - \beta) = \frac{\mathbf{F}_f}{\mathbf{F}_c} \tag{4.25}$$

Isolating the angle of kinetic friction (ϕ) leads to

$$\phi = \tan^{-1}\left(\frac{\mathbf{F}_f}{\mathbf{F}_c}\right) + \beta \tag{4.26}$$

With the angle of kinetic friction (ϕ) computed from Eq. 4.26, the frictional force ($\mathbf{F}$) and the normal force ($\mathbf{N}$) can be expressed as

$$\mathbf{F} = \mathbf{R} \sin \phi \tag{4.27}$$

$$\mathbf{N} = \mathbf{R} \cos \phi \tag{4.28}$$

An examination of Fig. 4.12 indicates that the shear force ($\mathbf{F}_s$) and the normal force ($\mathbf{F}_n$) are trigonometrically related to the resultant force ($\mathbf{R}$) by the interior angle ($\alpha - \beta + \phi$). In evaluating this angle, it is noted that the rake angle (β) is known for a given tool, the angle of kinetic friction (ϕ) can be computed from Eq. 4.26, and the shear angle (α) can be determined from Eq. 2.22 if the cutting ratio ($R_a = t_1/t_2$) is known. For convenience, the following substitution can be made:

$$K = \alpha - \beta + \phi \tag{4.29}$$

where

$$\alpha = \tan^{-1}\left(\frac{r_a \cos \beta}{1 - r_a \sin \beta}\right)$$

and

$$\phi = \tan^{-1}\left(\frac{\mathbf{F}_f}{\mathbf{F}_c}\right) + \beta$$

Equation 4.29 can then be written as

$$K = \tan^{-1}\left(\frac{r_a \cos\beta}{1 - r_a \sin\beta}\right) + \tan^{-1}\left(\frac{\mathbf{F}_f}{\mathbf{F}_c}\right) \tag{4.30}$$

With the angle K available in the form of Eq. 4.30, the shear force ($\mathbf{F}_s$) and the force normal to the shear plane ($\mathbf{F}_n$) can be expressed in terms of the resultant force as

$$\mathbf{F}_s = \mathbf{R}\cos K \tag{4.31}$$

$$\mathbf{F}_n = \mathbf{R}\sin K \tag{4.32}$$

The determination of the value of the individual component forces illustrated in Fig. 4.12 is presented with the following example.

EXAMPLE 4.7. A metal cutting operation is in the process of being evaluated with respect to the component forces shown in Fig. 4.12. Dynamometer readings indicate that the cutting force ($\mathbf{F}_c$) is equal to 300 lb (1334 N) and the feed force ($\mathbf{F}_f$) is equal to 125 lb (556 N). The rake angle on the tool (β) is 10°. A measurement of the chip thickness (t_2) indicates it is equal to 0.0343 in. (0.871 mm). The feed setting (t_1) of the machine is 0.015 in. (0.381 mm).

From these data it is desired to determine:

(a) The resultant force (**R**).
(b) The shear angle (α).
(c) The angle of kinetic friction (ϕ).
(d) The frictional force (**F**).
(e) The normal force (**N**).
(f) The angle K.
(g) The shear force ($\mathbf{F}_s$).
(h) The force normal to the shear plane ($\mathbf{F}_n$).

Solving for the resultant force by substituting into Eq. 4.24 yields

$$\begin{aligned}\mathbf{R} &= \sqrt{(\mathbf{F}_c)^2 + (\mathbf{F}_f)^2} \\ &= \sqrt{(300)^2 + (125)^2} \\ &= 325 \text{ lb } (1445.7 \text{ N})\end{aligned}$$

Solving for the shear angle by substituting into Eq. 2.22 yields

$$\tan\alpha = \frac{r_a \cos\beta}{1 - r_a \sin\beta}, \qquad \text{where } r_a = t_1/t_2$$

$$= \frac{0.4373(0.9848)}{1 - 0.4373(0.1736)}$$

$$\alpha = 25°$$

Solving for the angle of kinetic friction by substituting into Eq. 4.25 yields

$$\begin{aligned} \phi &= \tan^{-1}\left(\frac{\mathbf{F}_f}{\mathbf{F}_c}\right) + \beta \\ &= \tan^{-1}\left(\frac{125}{300}\right) + 10^\circ \\ &= 32.62^\circ \end{aligned}$$

Solving for the frictional force and the normal force by substituting into Eq. 4.27 and 4.28 yields

$$\begin{aligned} F &= \mathbf{R}\sin\phi \\ &= 325\sin 32.62^\circ \\ &= 175 \text{ lb } (778.4 \text{ N}) \end{aligned}$$

and

$$\begin{aligned} N &= \mathbf{R}\cos\phi \\ &= 325\cos 32.62^\circ \\ &= 274 \text{ lb } (1219 \text{ N}) \end{aligned}$$

Solving for the angle K by substituting into Eq. 4.30 yields

$$\begin{aligned} K &= \tan^{-1}\left(\frac{r_a\cos\beta}{1 - r_a\sin\beta}\right) + \tan^{-1}\left(\frac{\mathbf{F}_f}{\mathbf{F}_c}\right) \\ &= \tan^{-1}\left(\frac{0.4373(0.9848)}{1 - 0.4373(0.1736)}\right) + \tan^{-1}\left(\frac{125}{300}\right) \\ &= \tan^{-1}0.466 + \tan^{-1}0.4167 \\ &= 24.99^\circ + 22.62^\circ \\ &= 47.61^\circ \end{aligned}$$

Solving for the shear force and the force normal to the shear plane by substituting into Eqs. 4.31 and 4.32 yields

$$\begin{aligned} \mathbf{F}_s &= \mathbf{R}\cos K \\ &= 325\cos 47.61^\circ \\ &= 219 \text{ lb } (974 \text{ N}) \end{aligned}$$

and

$$\begin{aligned} \mathbf{F}_n &= 325\sin 47.61^\circ \\ &= 240 \text{ lb } (1068 \text{ N}) \end{aligned}$$

Table 4.1. Listing of Example of Calculations

Given Data					Calculated Values							
F_c	F_f	β	t_1	t_2	R	α	ϕ	F	N	K	F_s	F_n
300 lb	125 lb	10°	0.015 in.	0.0343 in	325 lb	25°	32.62°	175 lb	274 lb	47.61°	219 lb	240 lb

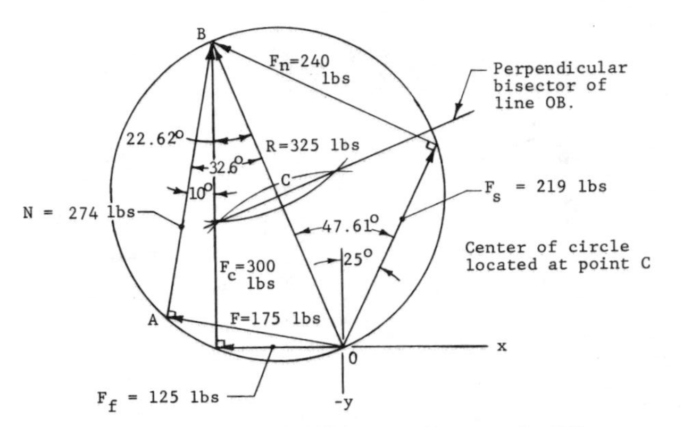

Fig. 4.14. Graphical layout of example 4.7.

The results of example 4.7 are tabulated in Table 4.1. A second method by which example 4.7 can be solved involves a graphical layout. By applying an appropriate scale to the cutting force ($\mathbf{F}_c$) and the feed force ($\mathbf{F}_f$), they can conveniently be drawn at right angles to each other. The hypotenuse of a right triangle having $\mathbf{F}_s$ and $\mathbf{F}_f$ as sides is a measure of the resultant force. By measuring the length of **R** and multiplying the length by the force scale, the value of **R** can be determined. Figure 4.14 represents a scaled graphical layout solution of example 4.7.

If one were inclined to graphically solve example 4.7, a convenient scale to use would be:

$$\text{Force scale:} \quad 1 \text{ in.} = 100 \text{ lb}$$

That is, 1 in. on the drawing represents a 100-lb force.

The midpoint of the resultant force **R** is obviously the center of the circle that encompasses the perpendicular force components. A perpendicular bisector of the line representing the resultant force **R** will conveniently locate the center of the circle. This can be drawn by setting a compass at a distance greater than half of the length of **R** and by swinging arcs from points O and B as shown in Fig. 4.14. Once this is accomplished, the circular arc can be drawn with its center located at point C.

Once the circle is drawn, the rake angle (β) set off at 10° enables the friction force (**F**) and the normal force (**N**) to be drawn. Finally, by setting off the shear angle (α), one can proceed to draw the shear force ($\mathbf{F}_s$) and the force normal to the shear plane ($\mathbf{F}_n$).

By measuring the length of each force vector and multiplying it by the force scale, it becomes possible to evaluate the individual force components. The angles between the force vectors can be measured directly from the diagram. This graphical technique provides not only a confirmation of the solutions calculated by formula as in example 4.7 but also clears the path to the solution of the same problem by a second method. For those who intuitively feel more comfortable with a graphical solution, laying out the force vectors to scale may prove less cumbersome than the analytical approach.

4.6 SHEAR STRESS

A measure of the shear stress acting on the shear plane of a metal cutting operation can be obtained in terms of the cutting force ($\mathbf{F}_c$), the feed force ($\mathbf{F}_f$), the feed (f), the depth of cut (d), and the shear angle (α). Figure 4.15 illustrates the shear force ($\mathbf{F}_s$) acting on the shear plane. The shear stress can be written as

$$S_s = \frac{\mathbf{F}_s}{A_s} \tag{4.33}$$

where S_s is the shear stress, $\mathbf{F}_s$ is the shear force, L_s is the shear length $= f/\sin\alpha$, A_s is the shear area $= L_s \times d$, and $A_s = f \times d/\sin\alpha$. By substituting, the shear stress can be expressed as

$$S_s = \frac{\mathbf{F}_s \sin\alpha}{f \times d} \tag{4.34}$$

Since dynamometer readings are given in terms of the cutting force ($\mathbf{F}_c$) and the feed force ($\mathbf{F}_f$), it is convenient to express the shear stress in these terms. Figure 4.16 illustrates the graphical relationship of the shear force ($\mathbf{F}_s$) with the cutting force ($\mathbf{F}_c$) and the feed force ($\mathbf{F}_f$). By projecting the cutting force ($\mathbf{F}_c$) and the feed force ($\mathbf{F}_f$) onto a plane parallel to the shear force ($\mathbf{F}_s$), it can be written that

$$\mathbf{F}_s = \mathbf{F}_c \cos\alpha - \mathbf{F}_f \sin\alpha \tag{4.35}$$

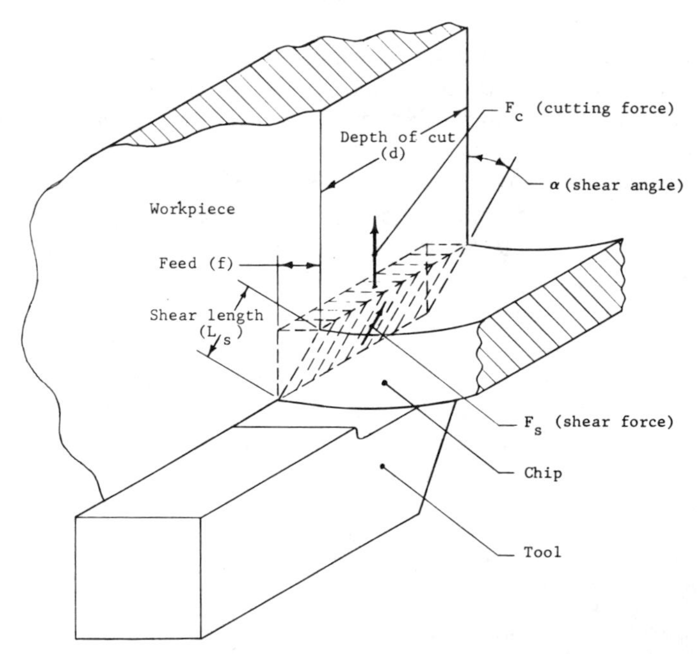

Fig. 4.15. Shear force acting on shear area, and cutting force acting on chip pressure area.

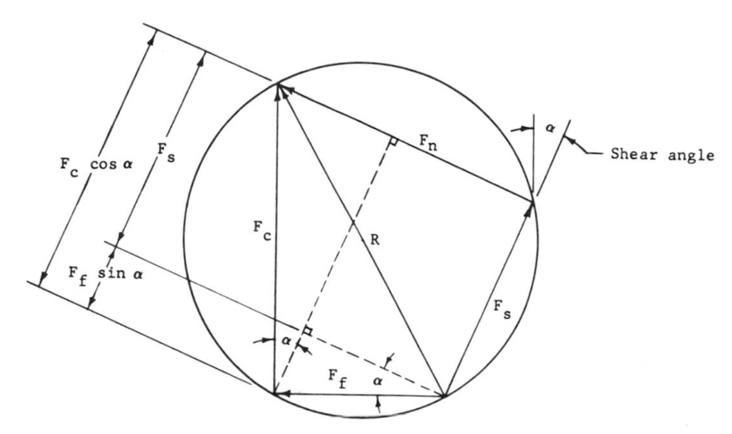

Fig. 4.16. Relationship between the shear force and the cutting and feed forces.

Substituting Eq. 4.35 into Eq. 4.34 yields

$$S_s = \frac{\mathbf{F}_c \cos\alpha \sin\alpha - \mathbf{F}_f \sin^2\alpha}{f \times d} \tag{4.36}$$

The following example illustrates how the shear stress can be calculated for a metal cutting operation by using dynamometer readings.

EXAMPLE 4.8. Using the data from example 4.7, where $\mathbf{F}_c = 300$ lb, $\mathbf{F}_f = 125$ lb, and $\alpha = 25°$, determine the shear stress for a turning operation. The feed (f) was set at 0.015 in./rev and the depth of cut (d) was set at 0.250 in.

Substituting into Eq. 4.34 yields

$$S_s = \frac{\mathbf{F}_c \cos\alpha \sin\alpha - \mathbf{F}_f \sin^2\alpha}{f \times d}$$

$$= \frac{300 \cos 25° \sin 25° - 125 \sin^2(25°)}{0.015 \times 0.250}$$

$$= 24{,}676 \text{ lb/in.}^2 \ (170 \times 10^6 \text{ Pa})$$

Another point of interest in analyzing stress is an appraisal of the chip pressure resistance. This is the ratio of the cutting force to the chip area perpendicular to the cutting force. This is illustrated in Fig. 4.17 as a segment removed from Fig. 4.15.

The chip pressure can be written as

$$P_c = \frac{\mathbf{F}_c}{f \times d} \tag{4.37}$$

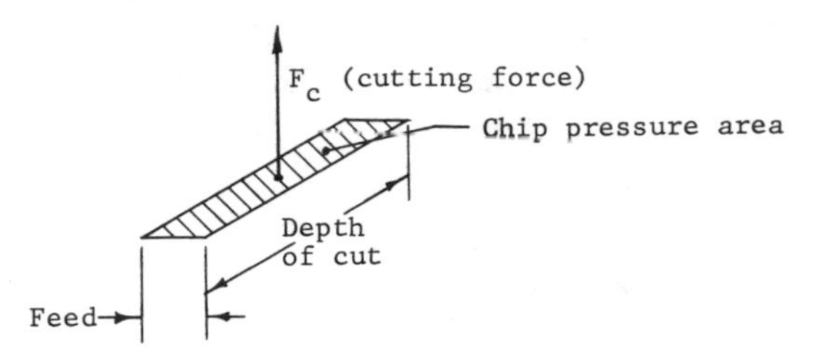

Fig. 4.17. Cutting force acting on chip pressure area.

where P_c is the chip pressure, $f \times d$ is the chip pressure area, and $\mathbf{F}_c$ is the cutting force. A calculation of the chip pressure is given with the following example, where a comparison is made between the chip pressure resistance and the shear stress for a metal cutting operation.

EXAMPLE 4.9. From the data collected in example 4.8, it is desired to calculate the chip pressure resistance and to compare it with the shear stress.

Substituting values into Eq. 4.37 yields

$$P_c = \frac{\mathbf{F}_c}{f \times d}$$

$$= \frac{300}{0.015 \times 0.250}$$

$$= 80{,}000 \text{ psi } (551.6 \times 10^6 \text{ Pa})$$

Solving for the ratio of the shear stress to the chip pressure resistance yields

$$R_{sc} = \frac{S_s}{P_c}$$

$$= \frac{24{,}676}{80{,}000}$$

$$= 0.308$$

4.7 INFLUENCE OF SHEAR ANGLE ON CUTTING FORCE

The cutting force ($\mathbf{F}_c$) that the tool exerts on the workpiece can usually be reduced in a metal cutting operation by increasing the rake angle (β) on the tool. This is a consequence of the general effect that

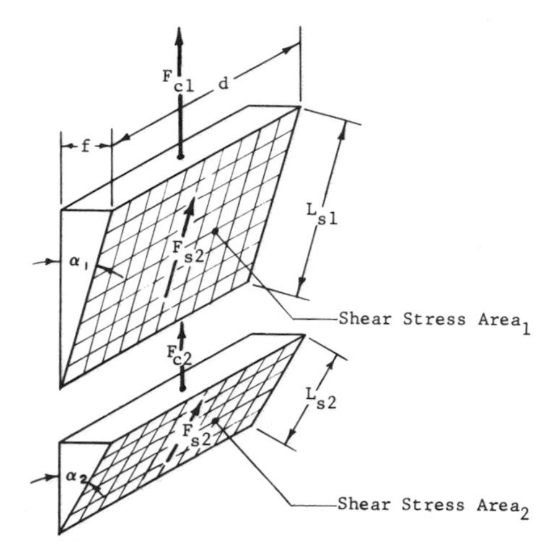

Fig. 4.18. Shear stress area reduction due to increase in shear angle.

an increase of the tool rake angle (β) has on the shear angle (α). Usually, a rake angle increase on the cutting tool will increase the shear angle on the work material. The result is a thinner chip and a reduction in the expenditure of energy. Figure 4.18 illustrates the effect of an increase in the shear angle on the shear area.

As can be seen in Fig. 4.18, the area ($f \times d$) over which the cutting force acts is equal to the feed (f) multiplied by the depth of cut (d). In the two cases shown, this area has the same value. However, an increase in the shear angle from α_1 to α_2, as shown, has a dramatic influence on the shear area over which the shear force acts. If the assumption is made that the shear strength of the material being cut is a true constant, that is, its value does not vary as the shear angle changes, then an interesting analysis can be conducted noting the influence of the shear angle on the cutting force.

Figure 4.19 illustrates the resultant force (**R**) resolved into two sets of perpendicular components. By projecting the shear force ($\mathbf{F}_s$) and the force normal to the shear plane ($\mathbf{F}_n$) onto the cutting force ($\mathbf{F}_c$), the following geometric relationship can be written:

$$\mathbf{F}_c = \mathbf{F}_s \cos\alpha + \mathbf{F}_n \sin\alpha \tag{4.38}$$

An examination of Fig. 4.19 reveals that the two sets of perpendicular components share a common hypotenuse (the resultant cutting force, **R**), which serves two right triangles. Taking the two right triangles, into account, it can be expressed that

$$\mathbf{R} = \frac{\mathbf{F}_c}{\cos(\phi - \beta)} = \frac{\mathbf{F}_s}{\cos(\alpha - \beta + \phi)} \tag{4.39}$$

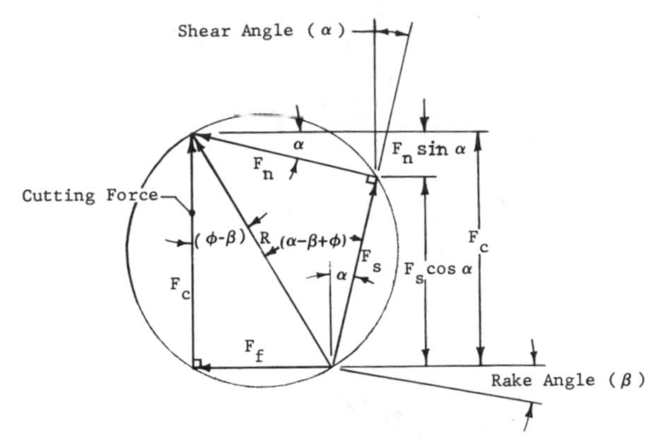

Fig. 4.19. Relationship between the cutting force and the shear force.

where α is the shear angle, β is the rake angle, ϕ is the angle of kinetic friction, ϕ is the $\tan^{-1}(\mathbf{F}/\mathbf{N}) = \tan^{-1}\mu$, and μ is the coefficient of friction. From Eq. 4.39 the cutting force ($\mathbf{F}_c$) can be expressed as

$$\mathbf{F}_c = \frac{\mathbf{F}_s \cos(\phi - \beta)}{\cos(\alpha - \beta + \phi)} \tag{4.40}$$

A simplified analysis of the influence of the shear angle on the forces acting between the tool and the work material will now be presented with the following assumptions imposed on the cutting process:

1. The shear strength of the material being cut remains constant.
2. The ratio between the shear force ($\mathbf{F}_s$) and the force normal to the shear plane ($\mathbf{F}_n$) remains constant.

In other words, the shear stress and the ratio $\mathbf{F}_n/\mathbf{F}_s$ are assumed not to vary during the chip forming operation as the shear angle changes due to different values being set on the rake angle of the tool. With the imposition of these two assumptions, the following example is scrutinized.

EXAMPLE 4.10. A turning operation is to be analyzed for the relationship between the shear angle and the cutting force. Experimental data reveal that the yield shear stress of the work material is 30,000 psi (206×10^6 Pa). In addition, the following measurements were taken:

Test	Rake Angle (β)	Shear Angle (α)
1	0°	9°
2	10°	13.5°
3	20°	18.0°
4	25°	20.0°

The feed was set at 0.005 in./rev (0.127 mm/rev) and the depth of cut was set at 0.250 in. (6.35 mm).

From these data and with the assumptions that the shear stress remains constant and that the ratio $\mathbf{F}_n/\mathbf{F}_s = 1$, it is desired to find the influence of the shear angle on the cutting force.

The shear area can be expressed as

$$A_s = \frac{f \times d}{\sin \alpha} \tag{1}$$

the shear force can be expressed as

$$\mathbf{F}_s = S_s \times A_s \tag{2}$$

and the cutting force can be expressed as

$$\mathbf{F}_c = \mathbf{F}_s \cos \alpha + \mathbf{F}_n \sin \alpha \tag{3}$$

where the assumption is made that $\mathbf{F}_n = \mathbf{F}_s$. With this substitution, Eq. 3 can be written as

$$\mathbf{F}_c = \mathbf{F}_s(\cos \alpha + \sin \alpha)$$

Substituting the given numerical values into Eqs. 1–3 gives the following results for the four tests:

Test	Shear Force, $\mathbf{F}_s$ (lb)	Shear Force, $\mathbf{F}_s$ (N)	Cutting Force, $\mathbf{F}_c$ (lb)	Cutting Force, $\mathbf{F}_c$ (N)
1	239.7	1066	274.25	1220
2	160.5	714	193.53	861
3	121.35	540	152.91	680
4	109.64	488	140.53	625

As can be seen in example 4.10, the shear angle, which has an affiliation with the tool rake angle, has, in turn, a strong influence on the cutting force. A graphically scaled diagram of the results of example 4.10 is shown in Fig. 4.20. A comparison of the cutting force in Test 1, 274.25 lb (1220 N), with the cutting force in Test 4, 140.53 lb (625 N), reveals that a 25° change in the rake angle ($\beta_4 - \beta_1$) results in a 48.8% reduction in the cutting force.

A point of interest in analyzing Fig. 4.20 is to see what effect the assumption that $\mathbf{F}_n/\mathbf{F}_s = 1$ had on the angle ($\phi - \beta$). It is obvious that if $\mathbf{F}_n/\mathbf{F}_s = 1$, then the angle ($\alpha - \beta + \phi$) must be equal to 45°. Since α and β are known angles, the angle of kinetic friction (ϕ) can be extracted in example 4.10 by writing that

$$\alpha - \beta + \phi = 45°$$

or

$$\phi = 45° - \alpha + \beta$$

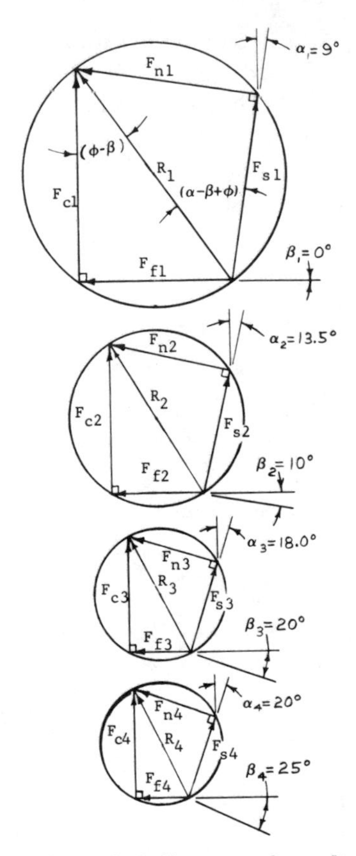

Fig. 4.20. Graphically scaled diagram of results of example 4.10.

Using the data for α and β from example 4.10 yields

$$\phi_1 = 45° - 9° + 0° = 36°$$

$$\phi_2 = 45° - 13.5° + 10° = 41.5°$$

$$\phi_3 = 45° - 18° + 20° = 47°$$

$$\phi_4 = 45° - 20° + 25° = 52°$$

The coefficient of friction can be extracted through the use of the angle ϕ by writing

$$\mu = \tan \phi = \mathbf{F}/\mathbf{N} \tag{4.6}$$

As a result,

$$\mu_1 = 0.727$$

$$\mu_2 = 0.885$$

$$\mu_3 = 1.072$$

$$\mu_4 = 1.280$$

As can be seen, to satisfy the equilibrium conditions as dictated by Fig. 4.20 with the imposition of the assumption that $\mathbf{F}_n/\mathbf{F}_s = 1$ and that the shear stress remains constant results in a wide fluctuation of the coefficient of friction. This leads to the questioning of the degree of reliability of the assumptions made. Although we are depending on pressure and temperature conditions that can lead up to a welding action between the tool and the chip resulting in a built-up edge on the tool, a large change in the coefficient of friction is possible. In a similar fashion, it can also be pointed out that the shear stress may vary as a function of the temperature of the work material, the rate of the shearing action, and the stress acting normal to the shear plane.

EXAMPLE 4.11. Calculate the expected cutting force for an operation where the shear stress is 35,000 psi (241×10^6 Pa) and the shear angle is 25°. The feed is set at 0.015 in. (0.381 mm) and the depth of cut is 0.125 in. (3.175 mm). Assume that the $\mathbf{F}_n/\mathbf{F}_s$ ratio is equal to 1.

Solving for shear area we get

$$A_s = \frac{f \times d}{\sin\alpha} = \frac{0.015 \times 0.125}{\sin 25^\circ}$$

$$= 0.0044 \text{ in.}^2 \ (0.1118 \text{ mm}^2)$$

The shear force is equal to

$$\mathbf{F}_s = S_s \times A_s = 35{,}000 \times 0.0044$$

$$= 154 \text{ lb } (685 \text{ N})$$

The cutting force is equal to

$$\mathbf{F}_c = \mathbf{F}_s \cos\alpha + \mathbf{F}_n \sin\alpha$$

If $\mathbf{F}_n = \mathbf{F}_s$, then

$$\mathbf{F}_c = \mathbf{F}_s(\cos\alpha + \sin\alpha) = 154(\cos 25^\circ + \sin 25^\circ)$$

$$= 204.7 \text{ lb } (910.3 \text{ N})$$

4.8 COEFFICIENT OF FRICTION

When two surfaces are in contact and one surface is moving relative to the other surface, as in the case of the chip and the tool in a metal cutting operation, a force resistant to the motion is generated and acts in a direction opposite to the relative velocity. This force is called the *frictional force*. It is a measure of the resisting force that the relative motion encounters. It can also be interpreted as a measure of the quality of the surface of a cutting tool. A tool that provides a low frictional force while the chip slides over it can be expected to expend less energy and to generate lower temperatures in the cutting operation.

Figure 4.21 illustrates the frictional force acting in a direction opposite to the relative sliding velocity. In Fig. 4.21(a) the velocity of the chip is viewed from the tool, that is, when viewed from the tool it is observed that the chip is moving in the direction shown. As can be seen, the frictional force that the tool exerts on the chip acts in a direction opposite to the chip motion. On the other hand, when viewed from the chip, it is observed that the tool is moving in the direction indicated in Fig. 4.21(b). In this case, the frictional force that the chip exerts on the tool has a tendency to try to move the tool along with the chip.

Newton's third law of mechanics states that when two bodies exert forces on each other, these forces are equal in magnitude, opposite in direction, and act in the same line of action. As a result, it can be written that

$$\begin{pmatrix}\text{Frictional force that}\\ \text{tool exerts on chip}\end{pmatrix} = -\begin{pmatrix}\text{Frictional force that}\\ \text{chip exerts on tool}\end{pmatrix}$$

or

$$\mathbf{F}_{t/c} = -\mathbf{F}_{c/t} \tag{4.41}$$

where the subscript t/c indicates tool relative to chip, and the subscript c/t indicates chip relative to tool.

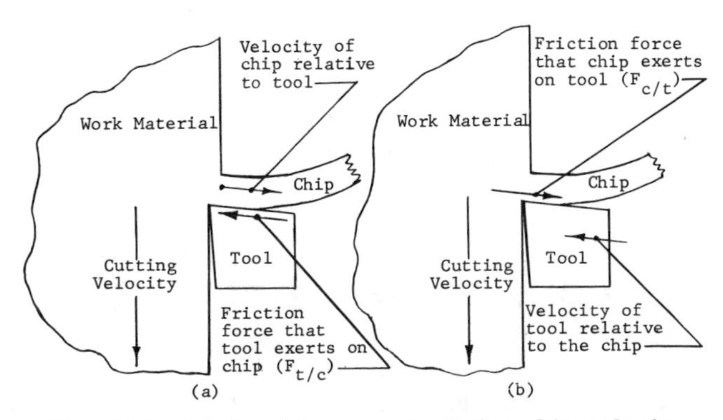

Fig. 4.21. Frictional force and relative chip velocity.

A measurement of the frictional-force-generating characteristic of surfaces in contact during sliding motion is known as the *coefficient of friction*. This is equal to the ratio of the friction force (**F**) to the force acting normal to the sliding surface (**N**). It can be expressed as

$$\mu = \frac{\mathbf{F}}{\mathbf{N}} \tag{4.5}$$

An examination of Fig. 4.4 reveals that trigonometrically this ratio can be expressed in terms of the angle ϕ as

$$\mu = \tan\phi = \frac{\mathbf{F}}{\mathbf{N}} \tag{4.6}$$

where ϕ is the angle of kinetic friction for sliding motion.

A test can be conducted to determine the value of the coefficient of friction in a metal cutting operation. However, using Eq. 4.5 is not convenient insofar as the frictional force (**F**) and the normal force (**N**) are difficult to measure directly. Usually in conducting a test of this type, a dynamometer is used, from which a measurement of the cutting force ($\mathbf{F}_c$) and the feed force ($\mathbf{F}_f$) can be determined. Figure 4.22 illustrates the action–reaction relationship between the work material and the cutting tool. $\mathbf{R}'$ represents the resultant force (action) that the work material exerts on the tool, whereas **R** represents the resultant force (reaction) that the tool exerts on the work material. As can be seen, the resultant is resolved into convenient components relating the cutting force ($\mathbf{F}'_c$) and the feed force ($\mathbf{F}'_f$) with the frictional force (**F**) and the normal force (**N**).

In order to evaluate the coefficient of friction directly from a dynamometer measurement of the cutting force ($\mathbf{F}_c$) and the feed force ($\mathbf{F}_f$),

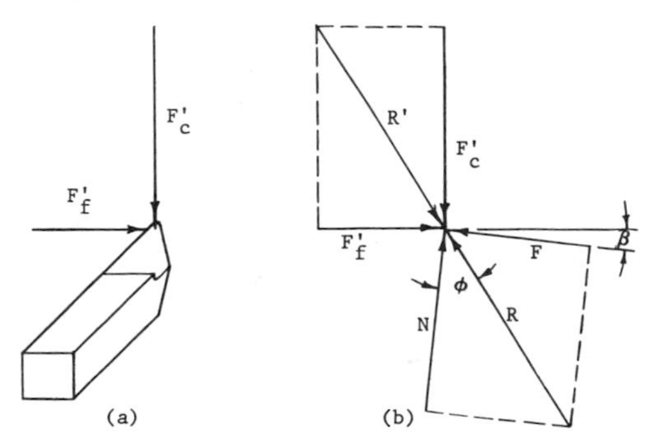

Fig. 4.22. Cutting and feed forces resolved into frictional and normal components.

expressions for the friction force (**F**) and the normal force (**N**) as functions of $\mathbf{F}_c$ and $\mathbf{F}_f$ are necessary. These can be conveniently attained from geometric projections, as shown in Fig. 4.23. The rake angle (β) is the trigonometric relationship that enables the projections to take place. By this technique, expressions can be derived that enable the coefficient of friction to be written in terms of the cutting force ($\mathbf{F}_c$), the feed force ($\mathbf{F}_f$), and the rake angle (β) of the tool.

To illustrate, by projecting the feed force ($\mathbf{F}_f$) and the cutting force ($\mathbf{F}_c$) onto the friction force (**F**) we can write the frictional force as

$$\mathbf{F} = \mathbf{F}_c \sin\beta + \mathbf{F}_f \cos\beta \tag{4.42}$$

In a similar fashion, the normal force can be written as

$$\mathbf{N} = \mathbf{F}_c \cos\beta - \mathbf{F}_f \sin\beta \tag{4.43}$$

Substituting into Eq. 4.6 yields

$$\mu = \tan\phi = \frac{\mathbf{F}_c \sin\beta + \mathbf{F}_f \cos\beta}{\mathbf{F}_c \cos\beta - \mathbf{F}_f \sin\beta}\left(\frac{\cos\beta}{\cos\beta}\right)$$

Multiplying by unity in the form of $\cos\beta/\cos\beta$ and noting that $\sin\beta/\cos\beta = \tan\beta$ simplifies the equation to

$$\mu = \tan\phi = \frac{\mathbf{F}_f + \mathbf{F}_c \tan\beta}{\mathbf{F}_c - \mathbf{F}_f \tan\beta} \tag{4.44}$$

The following numerical example demonstrates how the value of the coefficient of friction for a cutting operation can be extracted from dynamometer force measurements when the tool rake angle is known.

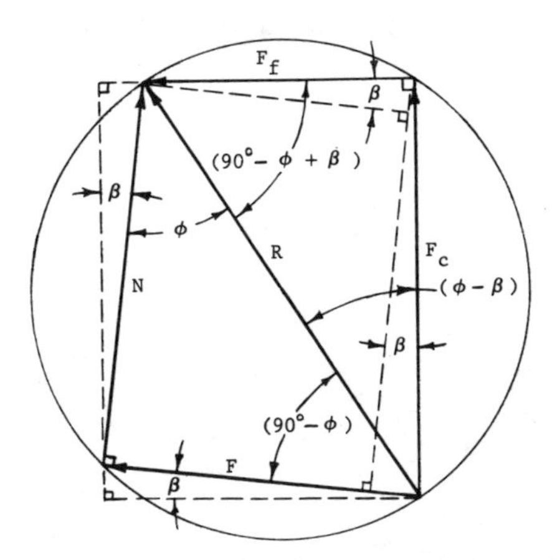

Fig. 4.23. Projection of cutting and feed forces in direction of frictional and normal forces.

EXAMPLE 4.12. Calculate the coefficient of friction for a metal cutting operation that has a measured cutting force of 310 lb (1379 N) and a feed force of 215 lb (956 N). The tool rake angle is 5°.

Substituting into Eq. 4.44 yields

$$\mu = \frac{215 + 310 \tan 5^\circ}{310 - 215 \tan 5^\circ}$$
$$= 0.8315$$

Another method by which the coefficient of friction can be determined from dynamometer readings and the value of the tool rake angle is through the use of the angle of kinetic friction (ϕ). An examination of Fig. 4.23 reveals that

$$\tan(\phi - \beta) = \frac{\mathbf{F}_f}{\mathbf{F}_c} \tag{4.45}$$

where

$$\phi = \left(\tan^{-1} \frac{\mathbf{F}_f}{\mathbf{F}_c}\right) + \beta \tag{4.46}$$

Substituting into Eq. 4.6 yields

$$\mu = \tan \phi = \tan\left[\left(\tan^{-1} \frac{\mathbf{F}_f}{\mathbf{F}_c}\right) + \beta\right] \tag{4.47}$$

EXAMPLE 4.13. Confirm the results of example 4.13 by using Eq. 4.47. Substituting numerical values into Eq. 4.46 results in

$$\mu = \tan[(\tan^{-1} \tfrac{215}{310}) + 5]$$
$$= \tan 39.74^\circ = 0.8315$$

As can be seen in example 4.13, the angle of kinetic friction is equal to 39.74°.

Of special interest in examining Fig. 4.23 in conjunction with equation 4.44 is the special case when the rake angle (β) is equal to zero. With $\beta = 0$, the angle $\phi - \beta$ in Eq. 4.45 becomes equal to ϕ and thus

$$\tan \phi = \frac{\mathbf{F}_f}{\mathbf{F}_c} = \mu \tag{4.48}$$

with $\beta = 0°$. In a fashion similar with $\beta = 0°$ in Eq. 4.44, the coefficient of friction can be written as

$$\mu = \tan \phi = \frac{\mathbf{F}_f}{\mathbf{F}_c} \tag{4.49}$$

where $\tan\beta = 0$. An appraisal of the effects of the condition $\beta = 0°$ in Fig. 4.23 indicates that the perpendicular components of **R** would form a rectangle where

$$\mathbf{N} = \mathbf{F}_c$$

and

$$\mathbf{F} = \mathbf{F}_f$$

As a result, with $\beta = 0°$,

$$\mu = \tan\phi = \frac{\mathbf{F}_f}{\mathbf{F}_c} = \frac{\mathbf{F}}{\mathbf{N}} \tag{4.50}$$

An important fact to note is that the coefficient of friction for metals in contact does not remain constant under conditions where the normal force, the contact temperature, and the relative velocity vary. In the metal cutting process, where high pressures exist between the surfaces, large deformations take place along the shear plane of the chip. This condition complicates the frictional relationship beyond a simple surface effect analysis. Physical properties of the metals in contact also have a major influence on the frictional force. At the points of contact between the chip and the tool, local welding adhesion and plastic flow add to the frictional resistance by requiring the shearing of the metallic junctions that are formed.

Test results of an experiment[1] carried out to measure the coefficient of friction applying Eq. 4.44 are shown in Table 4.2. A high-speed tool with a rake angle of 0° was used to machine a bar of cold-rolled steel. The depth of cut was set to 0.050 in. (1.27 mm) and the feed was set at 0.0041 in./rev (0.104 mm/rev). The cutting speed varied from 40 to 550 ft/min (12 to 167.6 m/min). The experimental data listed in Table 4.2 are plotted in Fig. 4.24.

An examination of the test results reveal that the coefficient of friction increased dramatically as a function of cutting speed for values below 350 ft/min (10.67 m/min). Above 400 ft/min (122 m/min), the experimental data indicate a constant coefficient of friction equivalent to 0.806. An intuitive evaluation of this response of coefficient of friction as a function of cutting speed could be justified by taking into account the complexity of the interaction of the chip flow over the tool. Work hardening of the chip as a result of plastic deformation, combined with high pressure and temperature conditions, can cause adhesion and

[1] Leslie F. Costa, Jr., "A Measurement of the Coefficient of Friction as a Function of Cutting Speed," Senior Design Project, Southeastern Massachusetts University, North Dartmouth, Massachusetts, 1977.

Table 4.2. Listing of Results of Coefficient of Friction Test

Cutting Speed		Dynamometer Readings				Coefficient of Friction
		F_c		F_f		
(ft/min)	(m/min)	(lb)	(N)	(lb)	(N)	
40	12.2	74	329	34	151	0.459
60	18.3	74	329	39	173	0.527
80	24.4	74	329	39	173	0.527
100	30.5	67	298	35	156	0.522
120	36.6	64	285	39	173	0.609
140	42.7	67	298	39	173	0.582
160	48.8	74	329	43	191	0.581
180	54.9	67	298	43	191	0.624
200	61.0	67	298	49	218	0.731
280	85.3	67	298	68	302	1.015
300	91.4	67	298	68	302	1.015
350	106.7	67	298	88	391	1.313
400	122.0	67	298	54	240	0.806
450	137.2	67	298	54	240	0.806
500	152.4	67	298	54	240	0.806
550	167.6	67	298	54	240	0.806

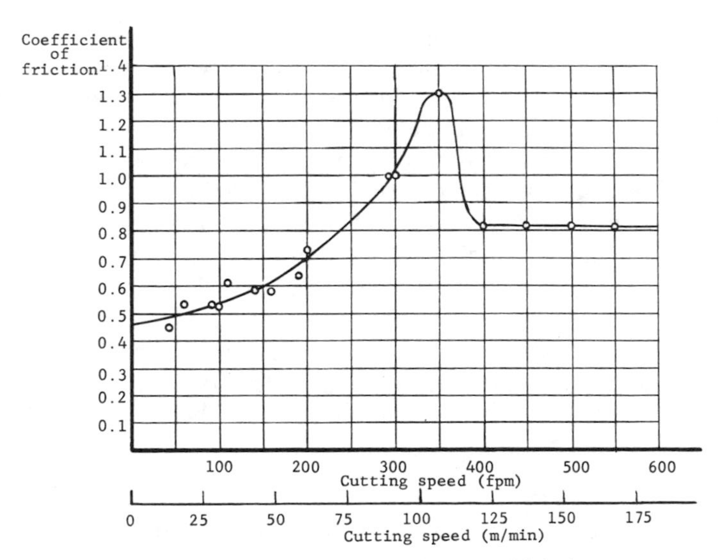

Fig. 4.24. Plot of results of coefficient of friction test.

welding of the points of contact between the chip and the tool. This state can lead to a built-up edge of the chip material on the tool. The result causes the chip formation to be formed by a ploughing of the softer chip material over the deposited harder chip material. What appears to be a high frictional force can be detected. This is, in reality, the force required to shear the chip material over the deposited built-up edge. The end product is a measured high coefficient of friction due to complications of internal chip behavior.

The drop in the coefficient of friction above the 350-ft/min (107-m/min) level can be rationalized by a removal of the built-up edge and a return to a simple sliding action between the chip and tool. This is contrasted with a complicated high-coefficient-of-friction metal shearing condition that may possibly exist in the 350-ft/min (107-m/min) range.

A point of interest would be the examination of a simple sliding friction case. Experimental results of sliding friction on dry surfaces of steel on steel reveal that, in general, the coefficient of friction declines as the sliding velocity between the surfaces increases. If this simple sliding friction case were imposed on a metal cutting operation, typical experimental data would be of the form indicated in Table 4.3. The data from Table 4.3 is graphically illustrated in Fig. 4.25.

A comparison of Fig. 4.24 with Fig. 4.25 indicates that there is a substantial difference in the shape of the curves. In reference sources, the value of the sliding coefficient of friction of hard steel on hard steel is usually given as about 0.42, whereas for mild steel on mild steel it is usually about 0.57. Focusing attention on the high values of the sliding coefficient of friction in metal cutting emphasizes the fact that there is a disparity when comparing these values with the results from the general laws of friction for clean, dry, and smooth surfaces.

In practice, the coefficient of friction associated with chip sliding has higher values than for ordinary sliding. The difference can be explained through the welding action that takes place between the chip and the tool. Many times, there is a welding of the chip to the tool and a mechanical interlocking of the surface asperites. This results in a deposit of a harder built-up edge on the tool requiring the

Table 4.3. Coefficient-of-Friction Sliding-Velocity Data

Sliding velocity						
(ft/min):	100	200	300	400	500	600
(m/min):	30.5	61.0	91.4	122	152	183
Coefficient of friction:	1.10	1.04	0.98	0.92	0.86	0.80

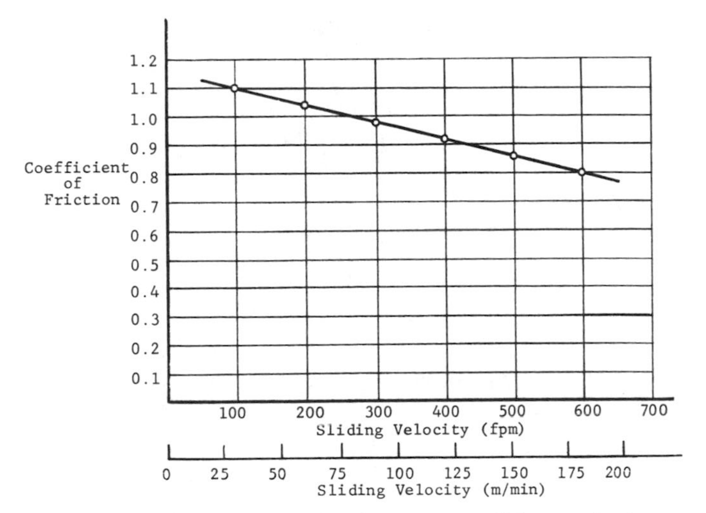

Fig. 4.25. Coefficient of friction versus sliding velocity.

ploughing of the deposited chip through the softer chip material. The result is dramatically unlike expectations from simple sliding friction analysis. In metal cutting, the coefficient of friction has a relationship with temperature, magnitude of sliding velocity, area of contact, and the applied normal load.

EXAMPLE 4.14. For a metal cutting operation with a tool rake angle of 0°, where the coefficient of friction is equal to 0.8 and the cutting force is measured to be 250 lb, calculate the magnitude of the feed force.

From Eq. 4.50,

$$\mathbf{F}_f = \mu \mathbf{F}_c \qquad \text{when } \beta = 0^\circ$$

Substituting yields

$$\mathbf{F}_f = 0.8(250) = 200 \text{ lb } (889.6 \text{ N})$$

4.9 VELOCITY RELATIONSHIPS

An examination of the metal cutting process reveals that there are three velocities that are of special significance. These are the cutting velocity ($\mathbf{V}_c$), the chip velocity ($\mathbf{V}_p$), and the shear velocity ($\mathbf{V}_s$). Figure 4.26 illustrates the different directions of these three velocities. The cutting velocity ($\mathbf{V}_c$) is the velocity of the work material relative to the

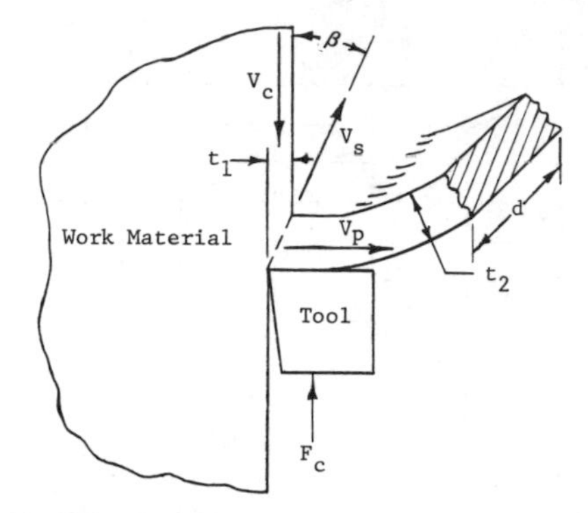

Fig. 4.26. Metal cutting velocities for stationary tool.

tool, and it has a direction parallel to the cutting force ($\mathbf{F}_c$). The chip velocity ($\mathbf{V}_p$) is the velocity of the chip relative to the tool, and it has a direction parallel to the tool face. The shear velocity ($\mathbf{V}_s$) is the velocity of the chip relative to the work material, and it has a direction parallel to the shear plane. If one were to observe the chip from the work material, it would appear as if the chip were moving in the direction of the shear velocity. As a result, it can be written that

$$\mathbf{V}_s = \mathbf{V}_p - \mathbf{V}_c \tag{4.51}$$

where $\mathbf{V}_s$ is the velocity of chip relation to work material, $\mathbf{V}_p$ is the velocity of chip relative to the tool, $\mathbf{V}_c$ is the cutting velocity or velocity of work material relative to the tool, and $\mathbf{V}_c = \mathbf{V}_{w/t} = -\mathbf{V}_{t/w}$ (w/t indicates work material relative to tool, and t/w indicates tool relative to work material).

Figure 4.27 is a graphical layout of Eq. 4.51, indicating that the shear velocity ($\mathbf{V}_s$) is vectorially equal to the chip velocity ($\mathbf{V}_p$) plus the velocity of the tool relative to the work material ($\mathbf{V}_{t/w}$).

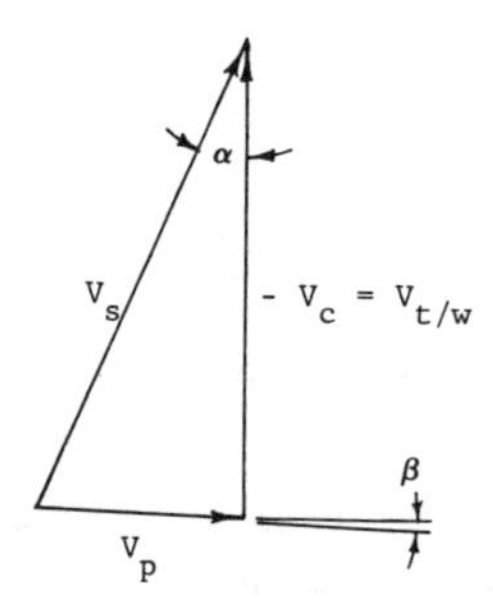

Fig. 4.27. Graphical layout of Eq. 4.51.

The relationship between the magnitudes of the cutting velocity and the chip velocity can be derived by noting that the volumetric quantity of material being cut per unit time must be accumulated in the chip. This can be expressed as

$$\text{Volume material} = \text{Volume chip}$$

or

$$\text{Feed} \times \text{Depth of cut} \times \text{Cutting velocity} = \text{Chip thickness} \times \text{Depth of cut} \times \text{Chip velocity}$$

or

$$t_1 \times d \times \mathbf{V}_c = t_2 \times d \times \mathbf{V}_p$$

from which

$$\mathbf{V}_p = \frac{t_1}{t_2}\mathbf{V}_c \tag{4.52}$$

EXAMPLE 4.15. Given a feed of 0.0041 in./rev (0.104 mm/rev), a cutting velocity of 300 ft/min (91.44 m/min), and a measured chip thickness of 0.016 in. (0.4064 mm), determine the chip velocity.

Substituting into Eq. 4.49 yields

$$\mathbf{V}_p = \frac{0.0041}{0.016}(300)$$

$$= 76.88 \text{ ft/min } (23.43 \text{ m/min})$$

Of interest in evaluating chip velocity ($\mathbf{V}_p$) is the observation that with the given cutting velocity ($\mathbf{V}_c$) two of the three vectors constituting Fig. 4.27 are available. If the angular orientation of the vectors is known, then the shear velocity as well as the shear angle (α) can be evaluated. The following example illustrates this technique.

EXAMPLE 4.16. Given the data from example 4.15, where the tool rake angle is 0°, solve for the shear velocity as well as the shear angle.

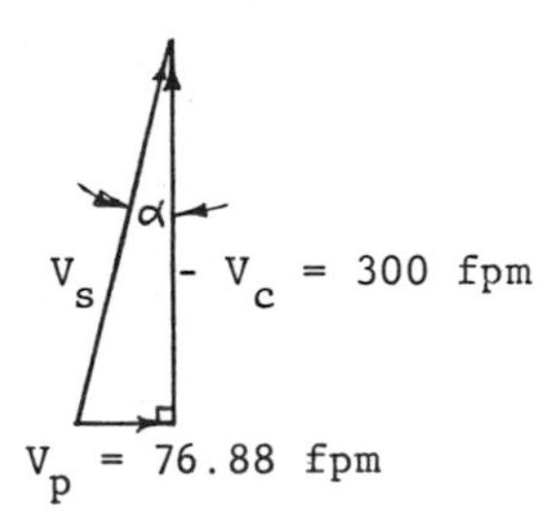

With $\beta = 0°$, the vector polygon constitutes a right triangle. Therefore, the magnitude of the shear velocity can be written as

$$\mathbf{V}_s = \sqrt{(\mathbf{V}_p)^2 + (-\mathbf{V}_c)^2}$$
$$= 309.7 \text{ ft/min } (94.4 \text{ m/min})$$

The shear angle (α) can be written as

$$\tan\alpha = \frac{\mathbf{V}_p}{\mathbf{V}_c} = \frac{76.88}{300} = 0.25627$$
$$\alpha = 14.37°$$

With the availability of Eq. 4.52, a chip thickness measurement can lead to the evaluation of the chip velocity for given machine settings of the feed (t_1) and the cutting velocity ($\mathbf{V}_c$). Table 4.4 lists values for the chip thickness as well as for the chip velocity for the coefficient of friction test specified in Table 4.2. The test involved turning cold-rolled steel with a high-speed tool that had a 0° rake angle. The feed was set at 0.0041 in./rev (0.104 mm/rev) and the depth of cut was 0.050 in. (1.27 mm). A plot of the chip thickness versus cutting velocity is given in Fig. 4.28.

As can be seen, the chip thickness (t_2) varies in conjunction with the changes in the coefficient of friction, which, in turn, responds to

Table 4.4. Chip Thickness and Sliding-Chip Velocity Values for Coefficient of Friction Test with Feed Setting of 0.0041 in./rev

Cutting Velocity, $\mathbf{V}_c$		Chip Thickness, t_2		Sliding Velocity, $\mathbf{V}_p$	
(ft/min)	(m/min)	(in.)	(mm)	(ft/min)	(m/min)
40	12.2	0.011	0.279	14.91	4.54
60	18.3	0.011	0.279	22.36	6.82
80	24.4	0.012	0.305	27.33	8.33
100	30.5	0.01125	0.286	36.44	11.11
120	36.6	0.012	0.305	41.00	12.50
140	42.7	0.012	0.305	47.83	14.58
160	48.8	0.011	0.279	59.64	18.18
180	54.9	0.012	0.305	61.50	18.75
200	61.0	0.013	0.330	63.08	19.23
280	85.3	0.0155	0.394	74.06	22.57
300	91.4	0.016	0.406	76.88	23.43
350	106.7	0.015	0.381	95.67	29.16
400	122.0	0.014	0.356	117.14	35.70
450	137.2	0.013	0.330	141.92	43.26
500	152.4	0.013	0.330	157.69	48.06
550	167.4	0.0125	0.318	180.40	54.99

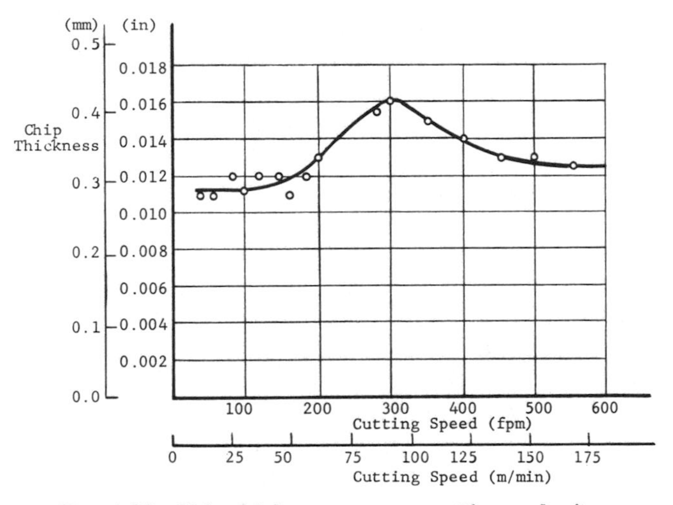

Fig. 4.28. Chip thickness versus cutting velocity.

the different settings of the cutting velocity. Of interest also is the relationship of the chip velocity to the cutting velocity. This is illustrated in Fig. 4.29.

The fluctuation in the slope of the chip-velocity–cutting-velocity curve indicates that the chip flow velocity as compared to the cutting velocity does not have a fixed linear relationship over the range of cutting velocity. The change is most evident between 200 and

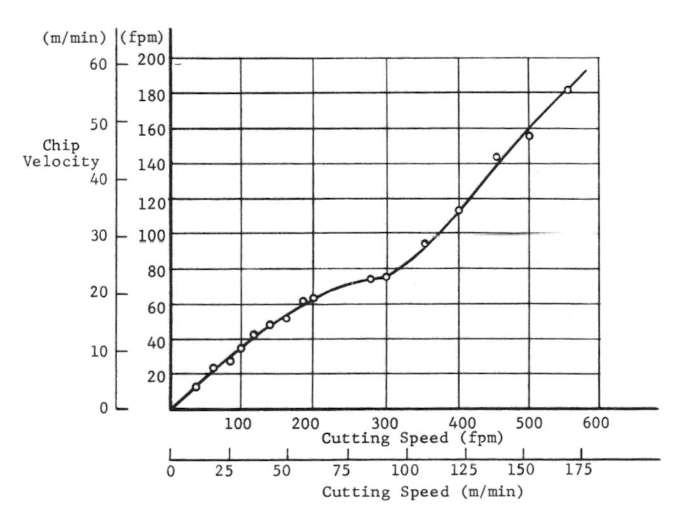

Fig. 4.29. Chip velocity versus cutting velocity.

300 ft/min, where the slope drops off significantly; this indicates an accumulating effect in the chip, which is confirmed by the corresponding chip thickness increase shown in Fig. 4.28.

The chip velocity ($\mathbf{V}_p$) and the shear velocity ($\mathbf{V}_s$) are related to the shear angle (α) and the rake angle (β), as well as with the cutting velocity ($\mathbf{V}_c$). Figure 4.30 shows these three velocities, along with the associated angles, for the case where the tool is considered stationary. This example is similar to the turning operation, where there is a tool feed velocity, but for the sake of the analysis it is so small on a relative scale that it can be neglected.

If the tool is considered to be fixed relative to the earth, and velocities relative to the earth are considered absolute, then the cutting velocity and the chip velocity can be considered to be absolute since they are both relative to the tool. An examination of the shear velocity indicates that relative to the tool it is zero since the shear plane remains in the same position when viewed from the tool. Note that the shear plane is the interface between the work material and the chip. If one places oneself on the work material and observes the motion of the chip, the relative velocity of the chip to the work material can be written in vectorial form as

$$\mathbf{V}_{p/c} = \mathbf{V}_p - \mathbf{V}_c = \mathbf{V}_s \tag{4.53}$$

If the cutting velocity ($\mathbf{V}_c$) is subtracted from the chip velocity ($\mathbf{V}_p$), the result is the velocity of the chip relative to the work material ($\mathbf{V}_{p/c}$). Figure 4.31 illustrates the vector polygon of Eq. 4.53, where $-\mathbf{V}_c$ is in the direction opposite to $\mathbf{V}_c$.

The relationships among the magnitudes of the velocities can be established by applying the law of sines for triangles. From Fig. 4.31,

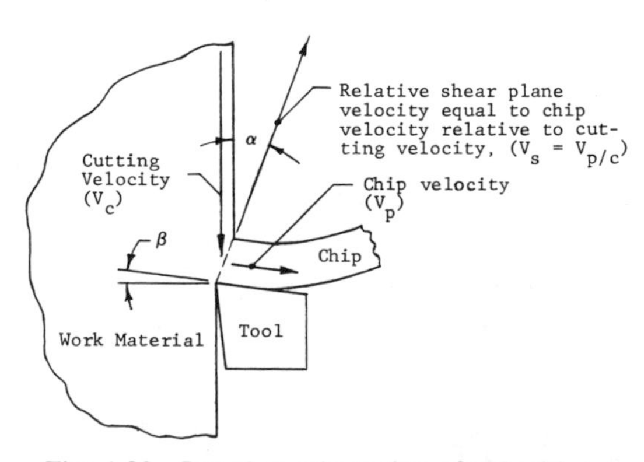

Fig. 4.30. Angular orientation of velocities.

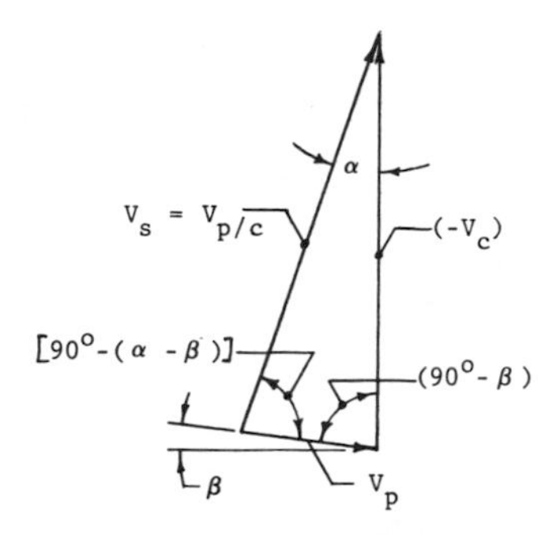

Fig. 4.31. Vector polygon representation of Eq. 4.53.

it can be written that

$$\frac{\sin \alpha}{\mathbf{V}_p} = \frac{\sin[90^\circ - (\alpha - \beta)]}{\mathbf{V}_c} = \frac{\sin(90^\circ - \beta)}{\mathbf{V}_{p/c}} \tag{4.54}$$

Using trigonometric identities where

$$\sin(90^\circ - \beta) = \cos \beta$$

and

$$\sin[90^\circ - (\alpha - \beta)] = \cos(\alpha - \beta)$$

and substituting into Eq. 4.54 yields

$$\frac{\sin \alpha}{\mathbf{V}_p} = \frac{\cos(\alpha - \beta)}{\mathbf{V}_c} = \frac{\cos \beta}{\mathbf{V}_s}, \qquad \text{where } \mathbf{V}_s = \mathbf{V}_{p/c} \tag{4.55}$$

Since the cutting velocity ($\mathbf{V}_c$) is set in a machining operation and can be readily determined, it is convenient to write the chip velocity ($\mathbf{V}_p$) and the shear velocity ($\mathbf{V}_s$) in terms of the cutting velocity ($\mathbf{V}_c$). Therefore, from Eq. 4.55,

$$\mathbf{V}_p = \frac{\sin \alpha}{\cos(\alpha - \beta)} \mathbf{V}_c \tag{4.56}$$

and

$$\mathbf{V}_s = \frac{\cos \beta}{\cos(\alpha - \beta)} \mathbf{V}_c \tag{4.57}$$

An examination of Eqs. 4.56 and 4.57 indicates that the chip velocity ($\mathbf{V}_p$) and the shear velocity ($\mathbf{V}_s$) are functions of the cutting velocity ($\mathbf{V}_c$), the shear angle (α), and the rake angle (β). In other words, if one is given $\mathbf{V}_c$, α, and β, then $\mathbf{V}_p$ and $\mathbf{V}_s$ can be evaluated. The following example demonstrates the application.

EXAMPLE 4.17. Confirm the results of examples 4.15 and 4.16 by using Eqs. 4.56 and 4.57, given the following data: cutting velocity, 300 ft/min (91.44 m/min); shear angle, 14.37°; and tool rake angle, 0°.

Substituting into Eq. 4.56 yields

$$\mathbf{V}_p = \frac{\sin 14.37^\circ}{\cos 14.37^\circ}(300) = 76.86 \text{ ft/min } (23.43 \text{ m/min})$$

Substituting into Eq. 4.57 yields

$$\mathbf{V}_s = \frac{\cos 0^\circ}{\cos 14.37^\circ}(300) = 309.7 \text{ ft/min } (94.4 \text{ m/min})$$

EXAMPLE 4.18. A test is being conducted to measure the chip velocity and the shear velocity for a metal cuttting operation. The cutting velocity is set at 300 ft/min (91.44 m/min) and the rake angle on the tool is 10°. A measurement of the chip thickness reveals that it is equal to 0.02286 in. (0.581 mm). What is the chip velocity and the shear velocity when the feed is set at 0.010 in. (0.254 mm)?

Solving for the shear angle by substituting into Eq. 2.22 yields

$$\tan\alpha = \frac{0.434 \cos 10^\circ}{1 - 0.4374 \sin 10^\circ} = 0.4663$$

$$\alpha = 25^\circ$$

Substituting into Eq. 4.56 yields

$$\mathbf{V}_p = \frac{\sin 25^\circ}{\cos 15^\circ}(300)$$

$$= 131.3 \text{ ft/min } (40.02 \text{ m/min})$$

Substituting into Eq. 4.57 yields

$$\mathbf{V}_s = \frac{\cos 10^\circ}{\cos 15^\circ}(300)$$

$$= 305.9 \text{ ft/min } (993.23 \text{ m/min})$$

Another method by which the shear velocity and the chip velocity can be determined is by means of a graphical layout. If the cutting velocity ($\mathbf{V}_c$) is drawn to scale and if the angles of α and β are known, then a vector polygon can be completed, yielding the magnitudes of $\mathbf{V}_p$ and $\mathbf{V}_s$. An illustration of this technique is given for the data from example 4.18. Figure 4.32 constitutes the graphical layout. The cutting velocity ($\mathbf{V}_c$) is drawn to scale and the known angles α and β are drawn. As can be seen in Fig. 4.32, the intersection of the known directions

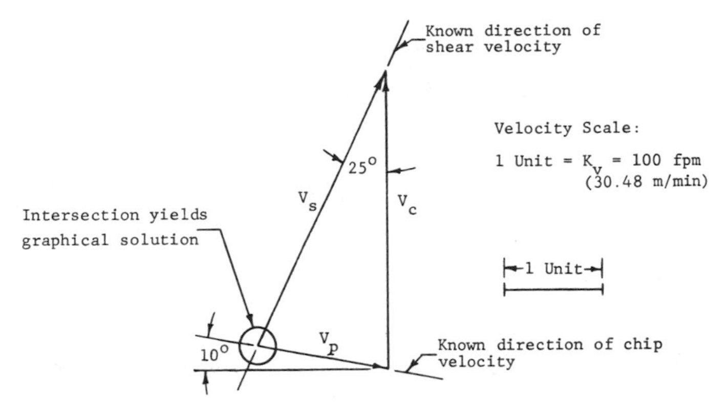

Fig. 4.32. Graphical solution of chip and shear velocities.

signified by α and β provide for a simultaneous graphical solution that yields the magnitudes of $\mathbf{V}_p$ and $\mathbf{V}_s$. Using the scale factor as shown, the velocities are confirmed to be equal to

$$\mathbf{V}_p = 131.3 \text{ ft/min } (40.02 \text{ m/min})$$

and

$$\mathbf{V}_s = 305.9 \text{ ft/min } (93.23 \text{ m/min})$$

Ordinarily, it is difficult to get exact accuracy when using a graphical layout. This is usually due to the precision required of the angular layout as well as the scaling measurement. However, with careful attention, an accurate range within a few percentage points can ordinarily be achieved with a graphical layout.

4.10 ENERGY ANALYSIS OF THE METAL CUTTING PROCESS

It is the objective of this section to show that in evaluating the energy expended in a metal cutting operation, the cutting speed and the cutting force are factors of prime importance. Work can be defined as the transference of energy by a process involving the motion of the point of application of a force. When there is movement against a resisting force, energy, a measure of work performed, is said to be used in providing the motion. Energy and work have the same units, given in terms of the product of the force and the displacement in the direction

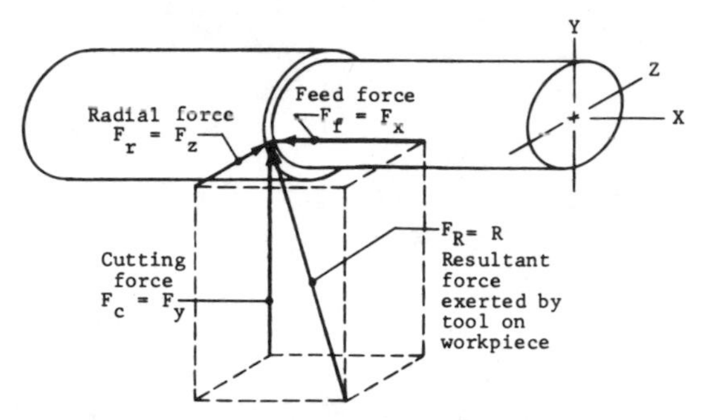

Fig. 4.33. Rectangular components of tool force acting on workpiece in turning operation.

of the force. In U.S. customary units, work is expressed in foot-pounds or inch-pounds, whereas in SI units work is expressed in newton-meters. The newton-meter, also known as a *joule*, is a unit of work:

$$\text{Energy} = \text{Work} = \text{Force} \times \text{Displacement}$$

$$E = W = \mathbf{F} \cdot \mathbf{D} \tag{4.58}$$

Figure 4.33 illustrates the rectangular components of the resultant tool force acting on the workpiece for a turning operation. Corresponding velocities are shown in Fig. 4.34, where the cutting velocity of the workpiece and the feed velocity of the tool are represented. The forces that the workpiece exerts on the tool are in the direction opposite to those that the tool exerts on the workpiece.

When an energy analysis (amount of work performed) of the three rectangular components of the cutting force ($\mathbf{F}_c$, $\mathbf{F}_f$, $\mathbf{F}_t$) is made, it becomes apparent that most of the energy of cutting is used by the cutting force ($\mathbf{F}_c$). This occurs not only because the cutting force is the largest of the component forces, but especially because the displacement in the direction of the cutting force is very large when compared

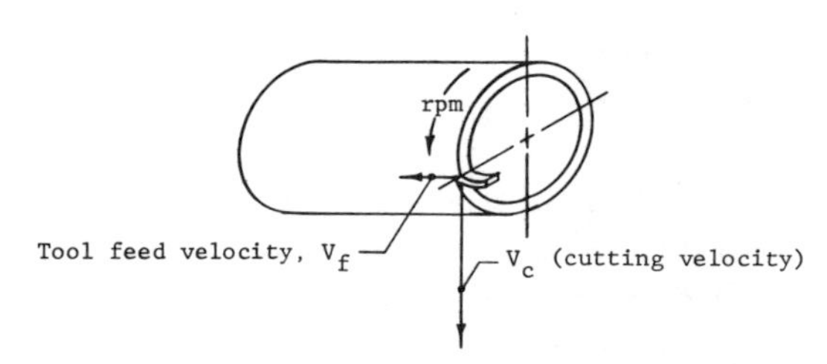

Fig. 4.34. Tool feed and cutting velocity.

to the displacement of the other component forces. As an example, the displacement for one revolution of the workpiece in the direction of the feed force is equal to the feed per revolution, whereas the displacement in the direction of the radial force is zero when turning a cylinder. When the displacement associated with the cutting force for one revolution of the workpiece is compared to the displacement of the feed force and the radial force, it is found to be relatively large. It is equivalent to the circumference of the workpiece.

EXAMPLE 4.19. It is desired to find the amount of work performed by each of the three components of the resultant force for one revolution of a turning operation. The cutting force measurements are $\mathbf{F}_c =$ 210 lb (934 N), $\mathbf{F}_f = 75$ lb (333.6 N), and $\mathbf{F}_r = 20$ lb (89 N). The diameter of the workpiece is 2 in. (50.8 mm), and it is being machined with a cutting velocity of 150 ft/min (45.72 m/min), a feed setting of 0.015 in./rev (0.381 mm/rev), and a depth of cut of 0.200 in. (5.08 mm).

Since the depth of cut is set at a fixed value, the displacement in the direction of the radial force is equal to zero. As a result, the work performed by the radial force can be written as

$$W_r = \mathbf{F}_r \cdot \mathbf{D}_r = 20 \times 0 = 0 \text{ in.-lb } (0 \text{ N-m})$$

The displacement for one revolution in the direction of the feed force is equal to the feed or 0.015 in. (0.381 mm). As a result, the work performed by the feed force can be written as

$$\begin{aligned} W_f &= \mathbf{F}_f \cdot \mathbf{D}_f = 75 \times 0.015 \\ &= 1.125 \text{ in.-lb } (0.127 \text{ N-m}) \end{aligned}$$

The displacement for one revolution in the direction of the cutting force can be written in terms of the circumference as

$$\mathbf{D}_c = \pi D = 3.14 \times 2 = 6.28 \text{ in. } (159.5 \text{ mm})$$

The work performed per revolution by the cutting force can now be written as

$$\begin{aligned} W_c &= \mathbf{F}_c \cdot \mathbf{D}_c = 210 \times 6.28 \\ &= 1318.8 \text{ in.-lb } (148.8 \text{ N-m}) \end{aligned}$$

A comparison of the magnitude of the work done by the component forces in example 4.19 indicates that practically all of the work was performed by the cutting force as a result of the affiliated large displacement in the direction of the cutting force. The percentage of work done by the feed force constitutes a very small portion of the total work performed. As a result, for practical purposes, the work performed by the feed force can be neglected. Consequently, the conclusion can be reached that the work done by the cutting force provides an accurate appraisal of the energy required for the cutting process. The following example delineates the percentage of energy used by each of the component forces in example 4.19.

EXAMPLE 4.20. Calculate the percentage of work performed by each of the three components of the resultant force in example 4.19.

Percentage of total work performed can be written as

$$\%\ W_{\text{comp}} = \frac{W_{\text{comp}}}{W_{\text{total}}} \times 100$$

where W_{comp} is the work performed by component force and W_{total} is the total work performed. Solving for the percentage of total work performed by the radial force yields

$$\%\ W_r = \frac{W_r}{W_c + W_f + W_r} \times 100 = \frac{0}{1318.8 + 1.125 + 0} \times 100$$
$$= 0$$

The percentage of total work performed by the feed force is written as

$$\%\ W_f = \frac{W_f}{W_c + W_f + W_r} \times 100 = \frac{1.125}{1319.93} \times 100$$
$$= 0.085\%$$

The percentage of total work performed by the cutting force is written as

$$\%\ W_c = \frac{W_c}{W_c + W_f + W_r} \times 100 = \frac{1318.8}{1319.925} \times 100$$
$$= 99.915\%$$

As can be seen from the results of example 4.20, 99.915% of the energy used in the cutting process described was expended by the cutting force, 0.085% was expended by the feed force, and 0.0% was expended by the radial force. From this analysis, the conclusion can be reached that for practical calculations, the energy expended to feed the tool into the workpiece and the energy expended by the radial force can be neglected.

The energy expended in the metal cutting process by multiple-edge cutting tools such as drills, end mills, milling cutters, hobs, and reamers can be analyzed in a fashion similar to that used for a single-edge turning tool. The key to the analysis is based on the reduction of the logical scheme to elementary fundamental principles. Figure 4.35 illustrates the resultant cutting force acting on one of the two cutting edges of a standard drill. Although in reality the cutting force is distributed over the length of the cutting edge, it can be represented in terms of a single concentrated force (**R**) acting at a distance (r_c) from the centerline of the drill. In Fig. 4.35, the resultant force is resolved into three convenient components, which are the cutting force ($\mathbf{F}_c$), the feed force ($\mathbf{F}_f$), and the radial force ($\mathbf{F}_r$). These forces are similar to those shown in Fig. 4.33 for the turning operation.

In analyzing the work performed by the radial component ($\mathbf{F}_r$) for the drilling operation, it can be concluded that this force travels in a circular path and that there is no displacement in the direction of this force, and, as a result, it does no work. The feed force ($\mathbf{F}_f$) in one revolution of the drill moves down a distance equivalent to the feed per revolution. The work it performs in one revolution is equal to the product of the feed force multiplied by the feed per revolution displacement. In analyzing the work performed by the cutting force ($\mathbf{F}_c$) in one revolution of the drill, the distance traveled by the cutting force

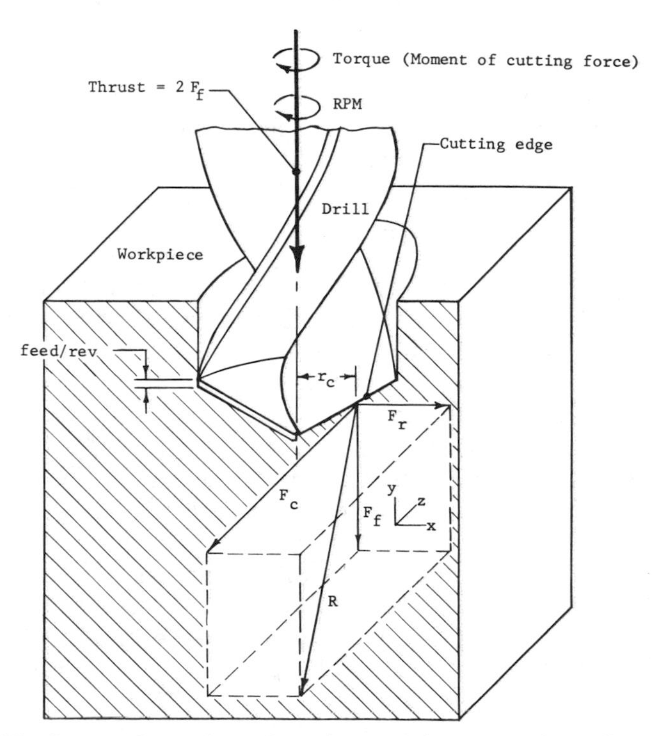

Fig. 4.35. Components of resultant force acting on cutting edge of drill.

must be taken into account. This displacement can be written as the circumference of the circle inscribed by the cutting force, or

$$\mathbf{D}_c = 2\pi r_c \tag{4.59}$$

The work of the cutting force for one revolution can now be written as

$$W_c = \mathbf{F}_c \cdot \mathbf{D}_c$$

or

$$W_c = \mathbf{F}_c(2\pi r_c) \tag{4.60}$$

The point described by r_c, at which the cutting force is concentrated, is an unknown quantity. Therefore, for the sake of convenience, it is more expedient to write Eq. 4.60 as

$$W_c = T(2\pi) \tag{4.61}$$

where $T = \mathbf{F}_c \times r_c$, which represents (1) the moment of the cutting force acting on the cutting edge and (2) the torque acting on drill; and 2π is the angle in radians through which the torque acts for one revolution of the drill.

As can be seen in the description of Eq. 4.61, the moment of the cutting force is equal to the product of the cutting force multiplied by the perpendicular distance extended from the cutting force to the center of rotation of the drill. In the example shown in Fig. 4.35, this distance is equal to r_c. It is noted that the moment of the cutting force is also referred to as the *torque*.

A more general expression of Eq. 4.61 is in the form

$$W = T \times \theta \tag{4.62}$$

where W is the work, T is the torque, and θ is the angle (in radians) through which the torque acts. The unit for work in Eq. 4.62 is the inch-pound or the foot-pound in the U.S. customary units and is the newton-meter in S.I. units. The unit of work, the newton-meter (N-m), is called a *joule*. It is noted that the torque has the same units as the work. The units are balanced on both sides of Eq. 4.62 because the unit for the angular displacement is given in radians, which, for this application, is not considered in the unit balance.

Figure 4.36 illustrates the measure of one radian, which, by definition, is a measure of a plane angle with its vertex at the center of a circle and subtended by an arc equal in length to the radius. Since the ratio between the radius and circumference of a circle is 2π, it can be stated that there are 2π radians in a circle. In degrees, one radian is equal to

$$1 \text{ rad} = \frac{360^\circ \text{ (degrees per circle)}}{2\pi\text{(radians per circle)}} = 57.3^\circ$$

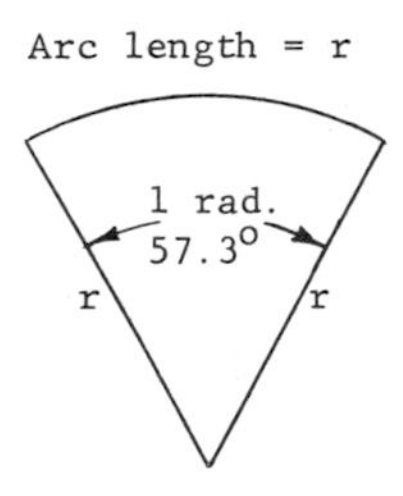

Fig. 4.36. The measure of one radian.

Equations for determining the torque and thrust of a standard drill, that is, one with a chisel length-to-diameter ratio of 0.18, have been derived by Shaw and Oxford.[2] The experiments that were conducted used SAE 3245 steel of 196–207 Bhn, with no cutting fluid, running at 20 ft/min. The equations are of the form

$$T = 123.54 \text{ Bhn } (f^{0.8} d^{1.8}) \tag{4.63}$$

and

$$T_{st} = 276.9 \text{ Bhn } (f^{0.8} d^{0.8}) + 3.124 \text{ Bhn } (d^2) \tag{4.64}$$

where T is the torque (in.-lb), T_{st} is the thrust (lb), Bhn stands for Brinell hardness number, f is the feed (in./rev), and d is the drill diameter (in.). An application of Eqs. 4.61 and 4.62 is illustrated in the following example.

EXAMPLE 4.21. Determine the torque and thrust for a $\frac{1}{2}$-in.-diameter drilling operation. The work material is steel with a Bhn of 200. The feed is set at 0.0097 in./rev and the cutting speed is 20 ft/min.

Solving for the torque by substituting into Eq. 4.63 yields

$$\begin{aligned} T &= 123.54(200)(0.0097)^{0.8}(0.5)^{1.8} \\ &= 173.7 \text{ in.-lb } (19.6 \text{ N-m}) \end{aligned}$$

Solving for the thrust by substituting into Eq. 4.64 yields

$$\begin{aligned} T_{st} &= 276.9(200)(0.0097)^{0.8}(0.5)^{0.8} + 3.124(200)(0.5)^2 \\ &= 935 \text{ lb } (4159 \text{ N}) \end{aligned}$$

With the availability of a means of determining the torque and the thrust, an analysis can now be made as to the distribution of energy between the cutting force and the feed force. It is noted that the cutting force is affiliated with the torque, whereas the feed force is affiliated with the thrust. The following example illustrates the analysis.

---

[2] M. C. Shaw, and C. J. Oxford, Jr., "On the Drilling of Metals 2—The Torque and Thrust in Drilling," *Trans. ASME*, Vol. 79, 1957, p. 147.

EXAMPLE 4.22. It is desired to find the amount of work (energy) performed in one revolution for the torque and the thrust for the drilling operation illustrated in example 4.21.

For one revolution, the torque acts through an angle of 360° or 2π rad. The thrust acts through a displacement equivalent to the feed, in this case 0.0097 in. in one revolution. From Eq. 4.62 the work performed by the torque acting through one revolution can be expressed as

$$W_c = T \times \theta$$

Substituting yields

$$\begin{aligned} W_c &= 173.7 \times 6.28 \\ &= 1090.9 \text{ in.-lb } (123.06 \text{ N-m}) \end{aligned}$$

With work being defined as the product of the force multiplied by the distance through which the force acts, the work performed by the thrust force per revolution of the drill can be written as

$$\begin{aligned} W_f &= T_{\text{st}} \times f \\ &= 935 \times 0.0097 \\ &= 9.07 \text{ in.-lb } (1.023 \text{ N-m}) \end{aligned}$$

As can be seen from the results of example 4.22, most of the energy of the drilling operation is expended by the cutting force. Similar to the turning example, less than 1% of the energy used to drill the parts is expended by the feed force, whereas more than 99% of the energy is used by the cutting force.

A point of interest would be a technique for extracting the magnitude of the cutting force acting on each of the cutting edges of the drill in examples 4.21 and 4.22. In order to do this, an assumption as to the value of r_c in Fig. 4.35 is necessary. If r_c is assumed to be equal to one-quarter of the diameter of the drill, then Fig. 4.37 represents the couple (two parallel forces of equal magnitude which act in opposite directions and are not collinear) that is generated by the cutting force acting on each cutting edge of the drill.

A couple consists of two forces having the same magnitude, parallel lines of action, and an opposite sense. It should be obvious that the summation of the forces of a couple is equal to zero. However, it is noted that the summation of the moments (force × distance) of a couple about any point is not equal to zero. Rather, the moment is equal to one of the forces of the couple multiplied by the perpendicular distance between the forces. As a result, it can be stated that the influence of

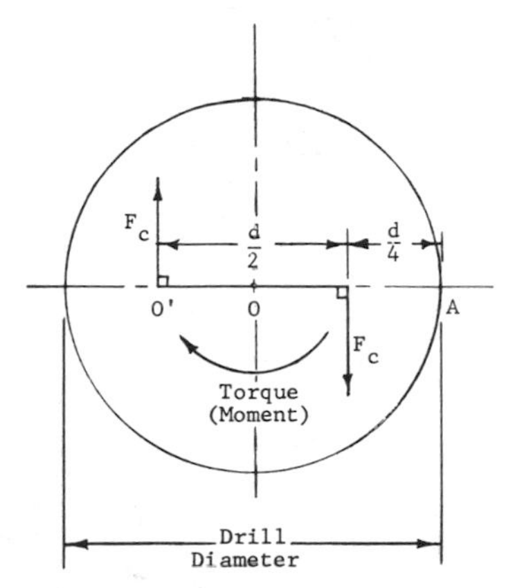

Fig. 4.37. Torque-generating couple exerted by drill on work material, with cutting force assumed to be concentrated at distance $r_c = \frac{1}{4}d$ from centerline.

a couple on a body is not zero. The two forces of a couple cause a torque (moment) to act on the body, providing a turning action.

An examination of Fig. 4.37 reveals that a summation of moments about point O is equal to

$$\curvearrowleft + \sum M_O = (\mathbf{F}_c \times \tfrac{1}{4}d) + (\mathbf{F}_c \times \tfrac{1}{4}d)$$

$$\curvearrowleft + \sum M_O = \mathbf{F}_c \times \tfrac{1}{2}d$$

Of interest is that a summation of moments about point 0′ or any other point (A) will yield the same results for a couple. To illustrate from Fig. 4.37,

$$\curvearrowleft + \sum M_{O'} = \mathbf{F}_c \times \tfrac{1}{2}d$$

$$\curvearrowleft + \sum M_A = (-\mathbf{F}_c \times \tfrac{1}{4}d) + (\mathbf{F}_c \times \tfrac{3}{4}d)$$

$$\curvearrowleft + \sum M_A = \mathbf{F}_c \times \tfrac{1}{2}d$$

EXAMPLE 4.23. With the assumption that the cutting force ($\mathbf{F}_c$) acting on a cutting edge of a drill is concentrated at a distance of $\frac{1}{4}d$ from the center of the drill, that is, $r_c = \frac{1}{4}d$, where d is the drill diameter, determine the cutting force from the data given in example 4.21.

Taking into account the moment of the cutting force couple yields

$$T = \mathbf{F}_c \times \tfrac{1}{2}d$$

$$\mathbf{F}_c = \frac{2T}{d} = \frac{2 \times 173.7}{0.5}$$

$$= 694.8 \text{ lb } (3090.6 \text{ N})$$

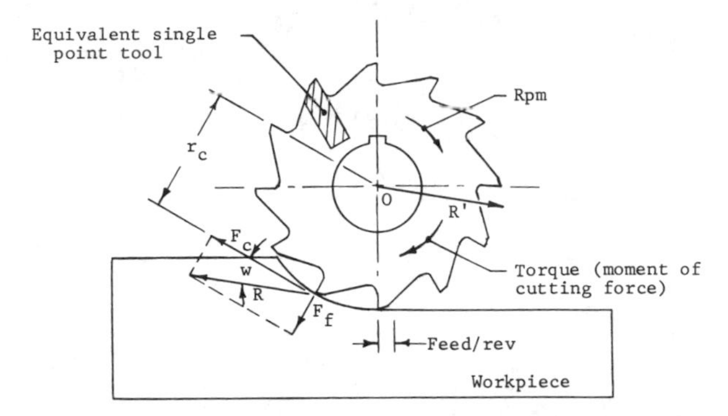

Fig. 4.38. Components of resultant force acting on cutting edge of milling cutter.

Figure 4.38 illustrates the cutting force exerted by a milling cutter on a workpiece. It is noted that the torque generated by the cutter is equal to the cutting force multiplied by the radius of the cutter. By taking a summation of moments about point O, it can be written that

$$\curvearrowleft + \sum M_O = T = \mathbf{F}_c \times r_c \tag{4.65}$$

The resultant force ($\mathbf{R}$), the vector summation of the cutting force ($\mathbf{F}_c$) and the feed force ($\mathbf{F}_f$), is also shown in Fig. 4.38. The reaction to the resultant force ($\mathbf{R}$) is labeled ($\mathbf{R}'$). This is the force that the cutter exerts on the arbor to which it is attached. Note that $\mathbf{R}$ and $\mathbf{R}'$ constitute a couple, and the moment of this couple is the same as that expressed in Eq. 4.65. It can be written as

$$T = \mathbf{R} \times r_c \cos w \tag{4.66}$$

where

$$\mathbf{R} \cos w = \mathbf{F}_c$$

4.11 POWER CONSUMPTION

Power is a convenient term that is used to measure the energy required to perform a metal cutting operation during a specific time period. By definition, power is equal to the energy (work) consumed per unit time or the rate at which work is performed. It can be expressed as

$$\text{Power} = \frac{\text{Work}}{\text{Time}}$$

$$P = \frac{W_c}{t} \qquad (4.67)$$

Since work is equal to the product of a force multiplied by the displacement in the direction of the force, Eq. 4.67 can be expressed as

$$P = \mathbf{F}_c \cdot \mathbf{V}_c \qquad (4.67a)$$

where $\mathbf{F}_c$ is the cutting force; and $\mathbf{V}_c = \mathbf{D}_c/t$, which represents velocity at cutting force or cutting speed. The unit for power in U.S. customary units is foot-pound per second, whereas in SI units it is newton-meter per second or joules per second.

A measure of power consumption in a metal cutting operation is important. Machine capacities, as an example, are determined in terms of how much energy per unit time can be delivered. If a high rate of cutting is desired, then a machine with the capacity to deliver the energy must be used. If the required power to perform a given operation exceeds the capacity of a machine tool, then obviously the machine will stall. The result is that machining rates can be limited if the capacity of the machine is below that demanded by the cutting process. Under these conditions, adjustments in machine settings must be made to match the machine capacity with the energy demands of the cutting process.

Figure 4.39 depicts the cutting velocity and cutting force acting in a turning operation. With the availability of these two values, power for an operation can be evaluated. The following example shows the method.

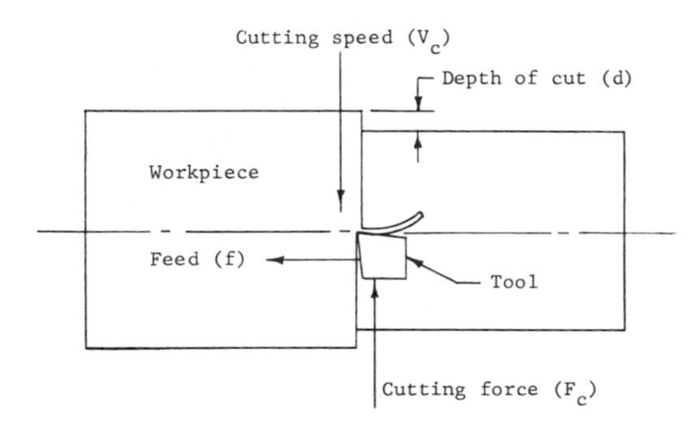

Fig. 4.39. Turning operation.

EXAMPLE 4.24. Calculate the power in foot-pounds per second (newton-meters per second) for a turning operation with a plain carbon steel workpiece that has a Brinell hardness number of 200 and a cutting velocity of 480 ft/min (146.3 m/min). A carbide tool is being used and machine settings are 0.020 in./rev (0.508 mm/rev) for the feed and 0.150 in. (3.81 mm) for the depth of cut. A measurement of the cutting force is found to be 1306 lb (5809.4 N).

Substituting numerical values into Eq. 4.67a enables the energy consumption per unit time to be calculated:

$$P = \mathbf{F}_c \cdot \mathbf{V}_c = 1306 \times 480$$
$$= 626{,}880 \text{ ft-lb/min} = 10{,}448 \text{ ft-lb/sec } (14{,}165.6 \text{ N-m/sec})$$

The units of power in example 4.24, foot-pounds per second (newton-meters per second), are affiliated with the definition of power in terms of the time rate at which work is being done. These units are obtained by dividing units of work by the unit of time. In U.S. customary units, the unit of power is known as *horsepower*, which is defined as

$$1 \text{ hp} = 550 \text{ ft-lb/sec}$$

In SI units, another measure of power is the *watt*, which is defined as

$$1 \text{ watt } (1 \text{ W}) = 1 \text{ joule/sec} = 1 \text{ newton-meter/sec}$$

The relationship between watts and horsepower are given as

$$1 \text{ hp} = 746 \text{ W}$$

Using the conversions for horsepower and watts, the result of example 4.24 can now be listed as

$$P = 10{,}448 \frac{\text{ft-lb}}{\text{sec}} \times \frac{1 \text{ hp-sec}}{550 \text{ ft-lb}}$$
$$= 19 \text{ hp } (14{,}171 \text{ W})$$
$$= 14.17 \text{ kW}$$

where 1 kW = 1000 W.

For a rotating tool, it is convenient to write the horsepower in terms of torque and rpm. A direct expression using these terms is

$$\text{hp} = \frac{\text{T} \times \text{rpm}}{63{,}025} \tag{4.68}$$

where T is the torque (in.-lb) and rpm stands for revolutions per minute, and 63,025 in.-lb-rpm/hp is the units conversion factor. Figure 4.40

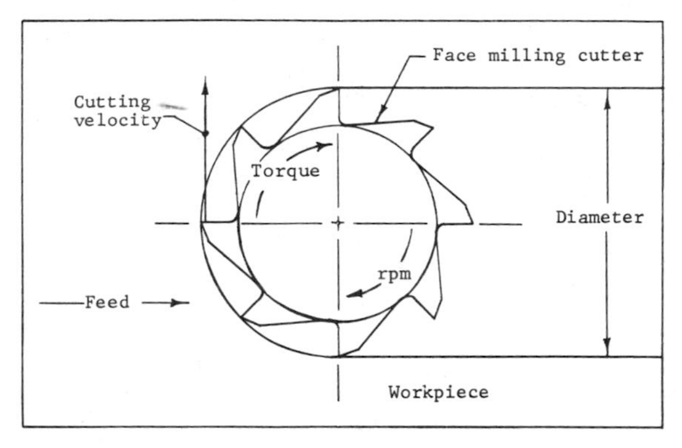

Fig. 4.40. Top view of face milling operation.

shows the torque and rpm acting on a face milling cutter. The following examples list basic calculations involving milling cutters.

EXAMPLE 4.25. A 2-in.(50.8-mm)-diameter end mill with six cutting edges (teeth) is programmed to rotate at 100 rpm. Calculate the cutting velocity and the feed per tooth for a workpiece feed of 1.75 in./min (44.45 mm/min).

From Eq. 2.1, substituting numerical values yields

$$\mathbf{V}_c = \frac{\text{rpm} \times \pi \times D}{12}$$

$$= \frac{100 \times 3.14 \times 2}{12}$$

$$= 52.3 \text{ ft/min } (15.95 \text{ m/min})$$

The advance of the end mill per revolution can be written as

$$\text{feed/rev} = \frac{\text{feed/min}}{\text{rev/min}} = \frac{1.75}{100}$$

$$= 0.0175 \text{ in./rev } (0.4445 \text{ mm/min})$$

Taking into account the six cutting edges yields

$$\text{feed/tooth} = \frac{\text{feed}}{\text{rev}} \times \frac{\text{rev}}{\text{cutting edges}}$$

$$= \frac{0.0175}{6}$$

$$= 0.0029 \text{ in. } (0.0737 \text{ mm})$$

EXAMPLE 4.26. The milling operation described in example 4.25 is set for a depth of cut of 0.500 in. (12.7 mm), and a torque measurement indicates a value of 100 lb-ft (135.6 N-m). The rpm is 100 and the work feed is 1.75 in./min (44.45 mm/min). From these data, calculate the horsepower required for the operation.

Substituting into Eq. 4.66 yields

$$\text{hp} = \frac{T \times \text{rpm}}{63{,}025} = \frac{100 \times 12 \times 100}{63{,}025}$$

$$= 1.9 \text{ horsepower } (1.42 \text{ kW})$$

Another way of expressing the horsepower required for a metal cutting operation is to write the equation in the form

$$\text{hp} = \frac{\mathbf{F}_c \cdot \mathbf{V}_c}{33{,}000} \tag{4.69}$$

where $\mathbf{F}_c$ is the cutting force (lb) and $\mathbf{V}_c$ is the cutting velocity (ft/min). Equation 4.69 can be derived directly from Eq. 4.68 by writing that the torque is equal to the average cutting force multiplied by the radius of the milling cutter. Substituting into Eq. 4.68 yields

$$\text{hp} = \frac{\mathbf{F}_c \times \frac{1}{2}D \times \text{rpm}}{63{,}025}$$

where

$$\mathbf{V}_c = \text{circumference} \times \text{rpm}$$

$$= \frac{\pi \times 2 \times \frac{1}{2}D \times \text{rpm}}{12}$$

and

$$\frac{D}{2} \times \text{rpm} = \frac{12\mathbf{V}_c}{2\pi}$$

Therefore

$$\text{hp} = \frac{\mathbf{F}_c(12\mathbf{V}_c/2\pi)}{63{,}025}$$

or

$$\text{hp} = \frac{\mathbf{F}_c \cdot \mathbf{V}_c}{33{,}000} \tag{4.69}$$

Equation 4.69 can be used for any metal cutting operation where the cutting force and cutting speed are known.

EXAMPLE 4.27. Confirm the results of example 4.26 by using Eq. 4.69.

The cutting force that generates the torque can be written as

$$\mathbf{F}_c = \frac{T}{D/2} = \frac{100 \times 12}{1}$$

$$= 1200 \text{ lb } (5337.9 \text{ N})$$

In example 4.25 the cutting velocity was determined to be equal to 52.3 ft/min (15.95 m/min). Substituting into Eq. 4.69 yields

$$\text{hp} = \frac{\mathbf{F}_c \cdot \mathbf{V}_c}{33{,}000} = \frac{1200 \times 52.3}{33{,}000}$$

$$= 1.9 \text{ horsepower } (1.42 \text{ kW})$$

The results of the turning example (example 4.24) can also be easily confirmed by Eq. 4.69, where it is noted that the constant 33,000 can be defined as

$$\frac{33{,}000 \text{ lb-ft}}{\text{min-hp}} = \frac{550 \text{ ft-lb}}{\text{sec-hp}} \times \frac{60 \text{ sec}}{\text{min}}$$

When accessibility to measurement of forces or torque is not available for the metal cutting operation, a technique by which energy and power requirements can be estimated is through the use of the term unit horsepower (unit power). This term provides for a measurement of the amount of power required to machine a metal at a specific volumetric rate. In U.S. customary units it is expressed as

$$U_p = \text{Unit horsepower (hp/in.}^3\text{/min)}$$

In SI units it is expressed as

$$U_p = \text{Unit power (kW/cm}^3\text{/sec)}$$

Since

$$1 \text{ hp} = 0.746 \text{ kW}$$

$$1 \text{ in.} = 2.54 \text{ cm}$$

$$1 \text{ in.}^3 = 16.387 \text{ cm}^3$$

and

$$1 \text{ min} = 60 \text{ sec}$$

then

$$1 \text{ hp/in.}^3\text{/min} = 0.746 \text{ kW}/16.387 \text{ cm}^3/60 \text{ sec}$$

$$1 \text{ hp/in.}^3\text{/min} = 2.73 \text{ kW/cm}^3\text{/sec}$$

Unit horsepower for various materials are found in reference sources and can be determined experimentally. As an example, if a wattmeter is attached to a machine tool and a measurement of the power demanded by the cutting operation is made, and the volumetric rate of metal removal is evaluated, then the unit power can be calculated. The following example illustrates the technique.

EXAMPLE 4.28. A wattmeter reading indicates that 14.174 kW are drawn by the required cut of a turning operation. The cutting speed is 480 ft/min (146.3 m/min), the feed is 0.020 in./rev (0.508 mm/rev), and the depth of cut is 0.150 in. (3.81 mm). From these data, calculate the unit horsepower (unit power), that is, the amount of power required to remove material at a specific rate.

A graphical description of the volumetric rate of machining is shown below.

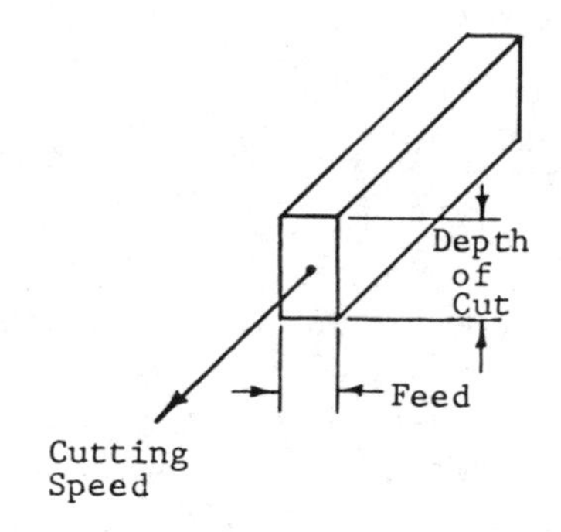

From Eq. 2.2 it can be written that

$$\begin{aligned} R_v &= 12\mathbf{V}_c \times f \times d \\ &= 12 \times 480 \times 0.02 \times 0.15 \\ &= 17.28 \text{ in.}^3/\text{min} \end{aligned}$$

The power required for this operation is

$$P = 14.174 \text{ kW}$$

or

$$\text{horsepower} = 19 \text{ hp}$$

Since the unit horsepower can be written as

$$U_p = \text{hp/in.}^3/\text{min} = \text{hp}_r/R_v$$

where hp_r is the required horsepower and R_v is the volumetric rate of machining, then by substituting

$$U_p = \frac{19}{17.28} = 1.1 \text{ hp/in.}^3/\text{min } (3.003 \text{ kW/cm}^3/\text{sec})$$

For example 4.28, in order to remove the material at a rate of 1 in.^{3}/min, 1.1 hp was required. The amount of energy (work) consumed to remove 1 in.^{3} (16.4 cm^3) can be calculated by substituting for horsepower and then multiplying by unity for unit balance in the unit horsepower equation. To illustrate:

$$U_p = \frac{1.1 \text{ hp-min}}{\text{in.}^3}$$

$$= \frac{1.1 \text{ hp} \times 550 \text{ ft-lb}}{\text{sec-hp}} \times \frac{\text{min}}{\text{in.}^3} \times \frac{60 \text{ sec}}{1 \text{ min}} \times \frac{12 \text{ in.}}{\text{ft}}$$

$$= \frac{435{,}000 \text{ in.-lb}}{\text{in.}^3}$$

As can be seen, the operation described requires 435,600 in-lb (44,676 joules) of energy to remove 1 in.^{3} (16,387 cm^3) of material.

Values of unit horsepower (unit power) vary with different materials and also vary with the hardness of a given material. The condition of the cutting tool can also have an influence. A dull tool ordinarily would require more energy to remove a specific amount of material than a sharp tool. In addition, the shape of the tool can also have an influence. A high side-rake-angle tool produces, on average, a lower chip thickness than a low-rake-angle tool. Consequently, this leads to a lower unit horsepower used in machining the same material. This is a result of less energy being absorbed in forming the thinner chip than in forming the thicker chip. The same argument can be extended to the sharp tool, which provides a cleaner chip and requires less energy than a tool that is dull. Figure 4.41 illustrates the wide distribution of

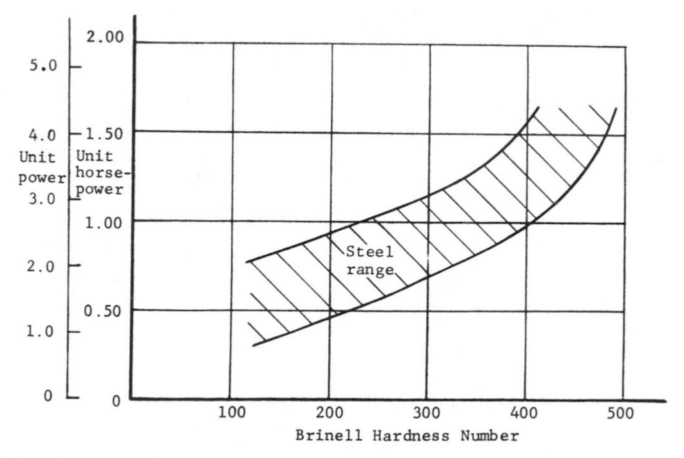

Fig. 4.41. Range of unit horsepower as a function of hardness for steel.

Table 4.5. Sample Values of Unit Horsepower

Material	Hardness (Bhn)	Unit Horsepower (hp/in.^{3}/min)	Unit Power (kW/cm^3/sec)
Steel	125	0.62	1.69
	180	0.75	2.05
	260	0.90	2.46
	430	1.50	4.10
Cast iron	150	0.4	1.09
	250	0.9	2.46
	270	1.2	3.28
Leaded brass	35	0.23	0.63
	75	0.26	0.71
	130	0.30	0.82
Copper	40	0.90	2.46
Aluminum	35	0.14	0.38
	90	0.16	0.44
	120	0.20	0.55
	150	0.24	0.66

unit horsepower (unit power) as a function of hardness for steel. The wide range for a given hardness reflects the effects of the constituents, condition, and heat treatment of the steel. Values at the lower portion of the range curve are for the free-cutting steel varieties, whereas the values at the higher portion of the range curve are for the high-carbon and alloy-steel varieties. Sample values of unit horsepower (unit power) for a variety of materials are shown in Table 4.5.

When using the unit horsepower (unit power) as a reference for calculations, the evaluation of the volumetric rate of metal removal becomes important. For the milling operation, the volumetric rate of metal removal can be written as

$$R_{vm} = f \times d \times w \tag{4.70}$$

where f is the feed (in./min), d is the depth of cut (in.), and w is the width of cut (in.). Figure 4.42 is a graphical representation of a volumetric rate of metal removal for a plain milling operation. A similar representation for the vertical milling case is shown in Fig. 4.43. In both cases, the volumetric rate of metal removal is symbolized by a right rectangular prismatic block of dimensions feed, depth of cut, and width of cut.

EXAMPLE 4.29. A 0.750-in.(19.05-mm)-diameter end mill is set to cut a 0.375-in.(9.525-mm)-deep slot in a low-carbon-steel workpiece that has a Brinell hardness number of 125. The high-speed-steel end mill is set

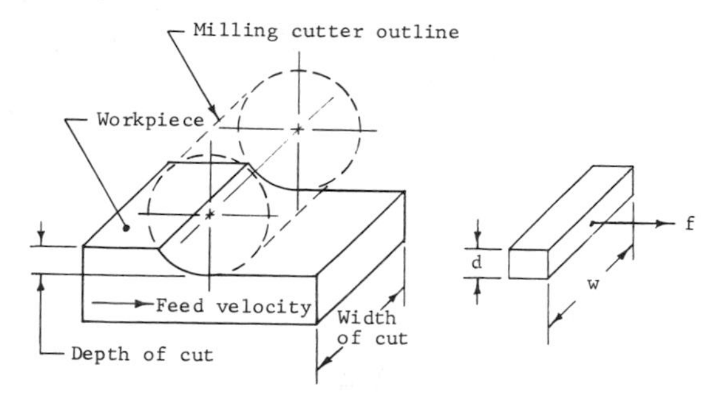

Fig. 4.42. Graphical representation of volumetric rate of metal removal for plain milling.

to run at 60 ft/min (18.29 m/min) with a recommended feed of 0.002 in. (0.0508 mm) per tooth. The end mill has four teeth (cutting edges).

From these data, calculate:

(a) The feed set on the machine tool.
(b) The volumetric rate of metal removal.
(c) The torque acting on the end mill by using a unit horsepower value from Table 4.5.
(d) The average cutting force acting on the end mill.

The feed per revolution of the end mill is

$$f_{\text{rev}} = 4 \times f_{\text{tooth}} = 0.008 \text{ in. } (0.203 \text{ mm})$$

The feed setting on the machine can be written as

$$f = f_{\text{rev}} \times \text{rpm}$$

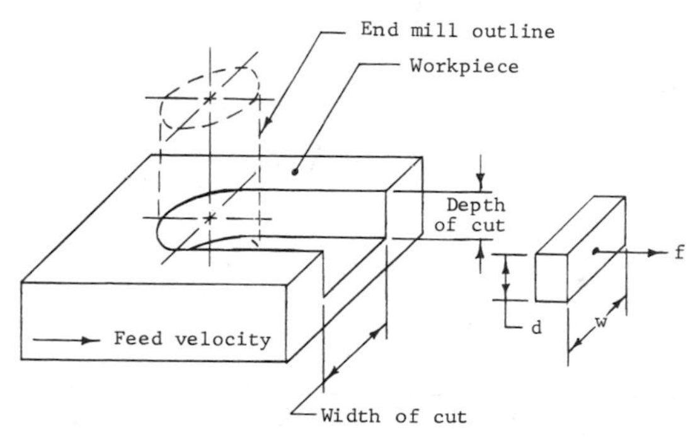

Fig. 4.43. Graphical representation of volumetric rate of metal removal for vertical milling cutter.

where

$$\text{rpm} = \frac{12\mathbf{V}_c}{\pi D} = \frac{12 \times 60}{3.14 \times 0.75} = 305.7$$

Thus,

$$\begin{aligned} f &= 0.008 \times 305.7 \\ &= 2.45 \text{ in./min } (62.23 \text{ mm/min}) \end{aligned}$$

The volumetric rate of metal removal can be evaluated by substitution in Eq. 4.70:

$$\begin{aligned} R_v &= f \times d \times w = 2.45 \times 0.375 \times 0.75 \\ &= 0.689 \text{ in.}^3\text{/min } (0.188 \text{ cm}^3\text{/sec}) \end{aligned}$$

From Table 4.5, for steel with a Bhn of 125 we have

$$U_p = 0.62 \text{ hp/in.}^3\text{/min}$$

Solving for the horsepower consumed in the operation, we obtain

$$\begin{aligned} \text{hp} &= U_p \times R_v = 0.62 \times 0.689 \\ &= 0.427 \text{ horsepower } (0.319 \text{ kW}) \end{aligned}$$

From Eq. 4.68,

$$\begin{aligned} T &= \frac{\text{hp} \times 63{,}025}{\text{rpm}} = \frac{0.427 \times 63{,}025}{305.7} \\ &= 88 \text{ lb-in. } (9.9 \text{ N-m}) \end{aligned}$$

The average cutting force can be written as

$$\mathbf{F}_c = \frac{T}{D/2} = \frac{88}{0.75/2} = 235 \text{ lb } (1044 \text{ N})$$

Of interest is that the average cutting force can also be extracted from Eq. 4.69 as

$$\begin{aligned} \mathbf{F}_c &= \frac{\text{hp} \times 33{,}000}{\mathbf{V}_c} = \frac{0.427 \times 33{,}000}{60} \\ &= 235 \text{ lb } (1044 \text{ N}) \end{aligned}$$

Figure 4.44 illustrates the volumetric rate of metal removal for a drilling operation. The volumetric rate can be expressed as

$$\begin{aligned} R_{\text{vd}} &= \text{Area of point cone} \times \text{Feed velocity} \\ &= A_{\text{pc}} \times f \end{aligned}$$

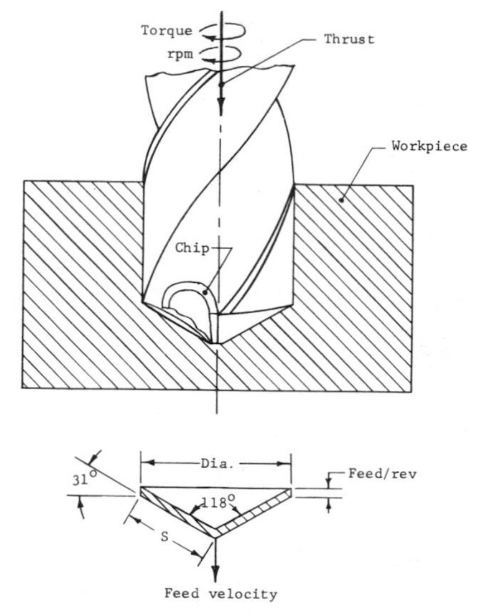

Fig. 4.44. Graphical representation of volumetric rate of metal removal for a drilling operation.

where

$$A_{pc} = \pi \frac{D}{2} \times S$$

$$\begin{aligned} S &= \text{Slant length of point cone} \\ &= \frac{\frac{1}{2}D}{\cos 31^\circ} \text{ (for standard drill)} \\ &= 0.5833 \times D \text{ (for standard drill)} \end{aligned}$$

Substituting for the area of the point cone generated by a drill yields

$$R_{vd} = \pi \frac{D}{2} \times S \times f \tag{4.71}$$

For a standard drill with a point angle of 118°,

$$S = S_{sd} = \frac{\frac{1}{2}D}{\cos 31^\circ} = 0.5833 \times D$$

and Eq. 4.71 can now be written as

$$R_{vd} = 0.9165 \times D^2 \times f \tag{4.72}$$

The following example demonstrates how the required torque for a drilling operation can be determined through the use of a reference unit horsepower (unit power) value.

EXAMPLE 4.30. A 0.500-in.(12.7-mm)-diameter high-speed standard drill is machining steel that has a Brinell hardness number of 260. The cutting speed is set at 40 ft/min (12.19 m/min) and the feed rate is 2 in./min (50.8 mm/min). Neglecting the effects of the chisel edge at the center of the drill point and using Table 4.5 as a reference, calculate the torque that is generated.

From Table 4.5,

$$U_p = 0.9 \text{ hp/in.}^3\text{/min } (2.46 \text{ kW/cm}^3\text{/sec})$$

The volumetric rate of metal removal can be determined from Eq. 4.72 as

$$R_{vd} = 0.9165D^2 \times f = 0.9165 \times (0.5)^2 \times 2$$
$$= 0.4583 \text{ in.}^3\text{/min}$$

Solving for the horsepower consumed in the operation, we obtain

$$\text{hp} = U_p \times R_{vd} = 0.9 \times 0.4583$$
$$= 0.4125 \text{ horsepower } (0.3075 \text{ kW})$$

The torque can now be determined by applying Eq. 4.68.

$$T = \frac{\text{hp} \times 63{,}025}{\text{rpm}}$$

where

$$\text{rpm} = \frac{12\mathbf{V}_c}{\pi \times D} = \frac{12 \times 40}{3.14 \times 0.5}$$
$$= 306 \text{ rpm}$$

Substituting yields

$$T = 84.95 \text{ lb-in. } (9.59 \text{ N-m})$$

The results of example 4.30 assume that the cutting edges of the drill act in a manner similar to that shown in Fig. 4.15. This is a close approximation when the effects of the chisel edge at the point of the drill are neglected. An example is the case where a pilot lead drill has been used and the drilling operation involves the enlargement of the hole. In this case, the chisel edge does no cutting.

However, in the ordinary case, the method of metal removal at the center of the drill is a complex one. It involves not only a cutting process but also an extrusion process. The influence of the action at the chisel edge can be significant. As a result, values of a simplified analysis as indicated in example 4.30 may need to be adjusted upward by a large factor (as large as 2 in some cases) to provide a more accurate appraisal.

For a turning operation similar to that shown in Fig. 4.39, the cutting force ($\mathbf{F}_c$) can be determined directly as a function of the feed (f), depth of cut (d), and the unit horsepower (unit power). This can be done by writing that

$$U_p = \frac{\text{hp}_c}{R_v} \tag{4.73}$$

where hp_c is the horsepower required at cutting tool,

$$\text{hp}_c = \frac{\mathbf{F}_c \cdot \mathbf{V}_c}{33{,}000}$$

and R_v is the volumetric rate,

$$R_v = 12f \times d \times \mathbf{V}_c \text{ (for turning)}$$

Substituting into Eq. 4.73 yields

$$U_p = \frac{(\mathbf{F}_c \cdot \mathbf{V}_c)/33{,}000}{12 \times f \times d \times \mathbf{V}_c}$$

or

$$U_p = \frac{\mathbf{F}_c}{396{,}000 \times f \times d}$$

Isolating the cutting force results in

$$\mathbf{F}_c = 396{,}000 \times f \times d \times U_p \tag{4.74}$$

As can be seen in Eq. 4.74, for a turning operation, if the feed, depth of cut, and unit horsepower are known, the cutting force can be directly evaluated. The constant 396,000 is for the U.S. customary unit system. The following example illustrates the application of Eq. 4.74.

EXAMPLE 4.31. It is desired to find the cutting force and power consumption for a turning operation. A medium-carbon-steel specimen with a Bhn of 180 is being machined by a C-6 carbide tool. The cutting velocity is set at 525 ft/min (160 m/min). The feed is 0.020 in./rev (0.508 mm/rev) and the depth of cut is 0.150 in. (3.81 mm). Reference sources reveal that the unit horse-power for the material is 0.75 hp/in.^{3}/min (2.048 kW/cm^3/sec).

Substituting numerical values into Eq. 4.74 yields

$$\begin{aligned}\mathbf{F}_c &= 396{,}000 \times f \times d \times U_p \\ &= 396{,}000(0.020 \times 0.150 \times 0.75) \\ &= 891 \text{ lb } (3963 \text{ N})\end{aligned}$$

The volumetric rate of machining for this operation is equal to

$$\begin{aligned}R_v &= 12 \times f \times d \times \mathbf{V}_c = 12 \times 0.020 \times 0.150 \times 525 \\ &= 18.9 \text{ in.}^3/\text{min } (5.16 \text{ cm}^3/\text{sec})\end{aligned}$$

The horsepower consumed by the cut can now be expressed as

$$\begin{aligned}\text{hp}_c &= U_p \times R_v = 0.75 \times 18.9 \\ &= 14.175 \text{ horsepower } (10.57 \text{ kW})\end{aligned}$$

4.12 MOTOR HORSEPOWER

In determining the power required to perform a specific metal cutting operation, two factors must be taken into consideration. The first of these is the efficiency of the power delivery system. The second is the amount of power necessary to drive a machine tool when no cutting is taking place, that is, when the machine is idling.

Efficiency is a measurement of the portion of energy received by a machine that is finally delivered to the cutting process in the form of useful work. It is expressed as a percentage. A machine that transforms into useful work 80% of the energy supplied to it is said to have an efficiency of 80%. In all machines that involve sliding motion, friction forces will do some work and the resultant output work will be smaller than the input work due to the frictional losses. A machine tool with a high efficiency has a large part of the energy supplied to it expended on the cutting process and only a small part wasted. Analytically, efficiency can be expressed as

$$\text{Eff} = \frac{\text{Output work}}{\text{Input work}} \tag{4.75}$$

Since power is the time rate of doing work expressed as

$$\text{Power} = \frac{\text{Work}}{\text{Time}}$$

then Eq. 4.75 can be written as

$$\text{Eff} = \frac{\text{Output power}}{\text{Input power}} \tag{4.76}$$

The following example shows how the overall efficiency of a machine tool can be measured for a metal cutting operation.

EXAMPLE 4.32. It is desired to measure the overall efficiency of a machine tool for a turning operation. A wattmeter is set up to measure the input power and a dynamometer is used to measure the cutting force. The workpiece has a 2.5-in. (63.5-mm) diameter and the machine is set to run at 230 rpm. The depth of cut is set at 0.125 in. (3.175 mm) and the feed is 0.006 in./rev (0.076 m/rev). The cutting force is found to be equal to 450 lb (2002 N) and the wattmeter reading during cutting indicates that 1.798 kW are being fed into the motor.

The input horsepower is

$$\mathrm{hp_{in}} = \frac{1.798}{0.746} = 2.41 \text{ hp}$$

The horsepower consumed at the cutting edge can be determined from Eq. 4.69:

$$\mathrm{hp} = \frac{\mathbf{F}_c \cdot \mathbf{V}_c}{33{,}000}$$

where

$$\mathbf{V}_c = \frac{\pi \times D}{12} \times \text{rpm}$$

$$= 150.5 \text{ ft/min } (45.87 \text{ m/min})$$

Substituting yields

$$\mathrm{hp} = \frac{450 \times 150.5}{33{,}000} = 2.052 \text{ horsepower}$$

Solving for efficiency yields

$$\mathrm{Eff} = \frac{\text{Output horsepower}}{\text{Input horsepower}} = \frac{2.052}{2.41}$$

$$= 0.851 \text{ or } 85.1\%$$

The 15% loss of energy in example 4.32 can readily be justified if an examination of the efficiency of individual machine elements is considered. If it is assumed that the input energy must be transmitted through the following system with the accompanying efficiencies,

Motor	0.97
Belt drive	0.95
One set of bearings	0.96
Second set of bearings	0.96

then the system efficiency can be written as

$$\begin{aligned}\text{Eff} &= \text{Eff(motor)} \times \text{Eff(belt)} \times \text{Eff(bearing 1)} \times \text{Eff(bearing 2)}\\ &= 0.97 \times 0.95 \times 0.96 \times 0.96\\ &= 0.849 = 84.9\%\end{aligned}$$

The second factor of consideration in determining the power required to perform a specific metal cutting operation on a machine tool is that of *tare horsepower*. This is the amount of horsepower that is required to overcome the machine's resistance to motion when the machine is running and cutting is not taking place. This is often referred to as the *idling horsepower*. The tare horsepower can be significant in cases where machine tools have gear trains with a large number of gears in contact. Bearings with high frictional forces can also increase tare horsepower as can machine elements such as belt drives. Contrasting high-tare-horsepower machine tools are those that possess anti-friction bearings, small numbers of gears in the gear train, and direct motor drives. These have high efficiencies and may possess relatively low tare-horsepower demands, and as a result have relatively high efficiencies.

A means of calculating the required horsepower for a motor on a machine tool is to take into consideration three factors that influence the motor horsepower. These can be expressed analytically as

$$\text{hp}_m = \frac{\text{hp}_c}{\text{Eff}} + \text{hp}_t \tag{4.77}$$

where hp_m is the motor horsepower, hp_c is the horsepower required at cut, hp_t is the tare horsepower, and Eff is the machine efficiency. The following examples give applications of Eq. 4.77.

EXAMPLE 4.33. A machine tool with a power efficiency of 80% and a tare horsepower of 0.56 hp is used for a turning operation. The material being cut is found to have a unit horsepower of 1.3. The cutting velocity is set at 175 ft/min (53.34 m/min), with a feed of 0.010 in./rev (0.254 mm/rev) and a depth of cut of 0.375 in. (9.525 mm). Determine the motor horsepower required for this operation.

The volumetric rate of metal removal is equal to

$$\begin{aligned}R_v &= 12 \times f \times d \times \mathbf{V}_c = 12 \times 0.010 \times 0.375 \times 175\\ &= 7.875 \text{ in.}^3/\text{min}\end{aligned}$$

From the unit horsepower, the horsepower required at the cut is equal to

$$\begin{aligned}\text{hp}_c &= U_p \times R_v = 1.3 \times 7.875\\ &= 10.24 \text{ horsepower}\end{aligned}$$

Substituting numerical values into Eq. 4.77 yields

$$hp_m = \frac{hp_c}{\text{Eff}} + hp_t = \frac{10.24}{0.80} + 0.56$$

$$= 13.36 \text{ horsepower } (9.966 \text{ kW})$$

EXAMPLE 4.34. The diagram shown below represents a 5-in.(127-mm)-diameter milling cutter machining a steel specimen that has a Bhn of 240 and a unit horsepower of 0.85. The milling cutter has eight teeth and the cutting velocity has been set at 200 ft/min (61 m/min). The depth of cut is 0.375 in. (9.525 mm) and the width of cut is 4 in. (101.5 mm). The feed rate is 0.004 in./tooth (0.106 mm/tooth). It is assumed that the milling takes place on a machine that has a power efficiency of 0.8 and a tare horsepower of 0.6.

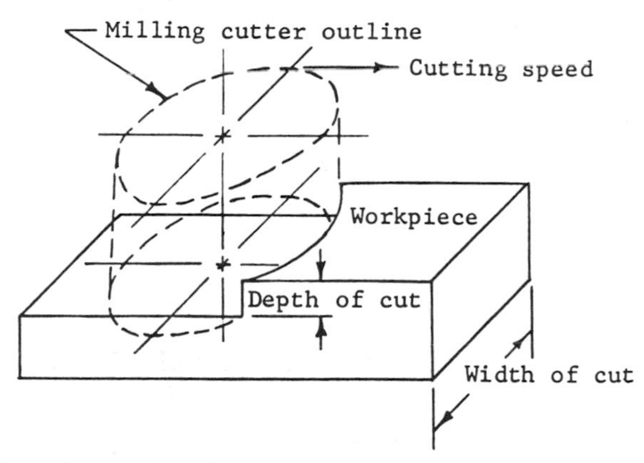

From this information determine:

(a) The rpm of the cutter.
(b) The feed rate of the workpiece.
(c) The volumetric rate of metal removal.
(d) The horsepower required at the cut.
(e) The required motor horsepower.
(f) The torque generated by the cut.

The rpm of the cutter can be written as

$$\text{rpm} = \frac{12 \times V_c}{\pi \times D} = \frac{12 \times 200}{3.14 \times 5} = 152.9 \text{ rpm}$$

The feed rate of the workpiece is

$$\begin{aligned} f_m &= \text{feed/tooth} \times \text{number of teeth/rev} \times \text{rpm} \\ &= 0.004 \times 8 \times 152.9 \\ &= 4.89 \text{ in./min} \end{aligned}$$

The volumetric rate of metal removal is equal to

$$R_{vm} = f_m \times d \times w = 4.89 \times 0.375 \times 4$$
$$= 7.335 \text{ in.}^3/\text{min } (2.002 \text{ cm}^3/\text{sec})$$

Taking into account the unit horsepower, the horsepower required at the cut can be written as

$$\text{hp}_c = U_p \times R_v = 0.85 \times 7.335$$
$$= 6.23 \text{ horsepower } (4.651 \text{ kW})$$

The required motor horsepower can be expressed as

$$\text{hp}_m = \frac{\text{hp}_c}{\text{Eff}} + \text{hp}_t = \frac{6.23}{0.8} + 0.6$$
$$= 8.39 \text{ horsepower } (6.26 \text{ kW})$$

The torque exerted on the milling cutter can be extracted from Eq. 4.68.

$$T = \frac{\text{hp}_c \times 63{,}025}{\text{rpm}} = \frac{6.23 \times 63{,}025}{152.9}$$
$$= 2568 \text{ lb-in. } (289.7 \text{ N-m})$$

4.13 EFFECTIVE TOOL CLEARANCE ANGLES

The angle that appears on a tool as a clearance or relief angle is influenced in an actual cutting operation by the rate of feed. Figure 4.45 shows the effect of the lead angle with regard to reducing the

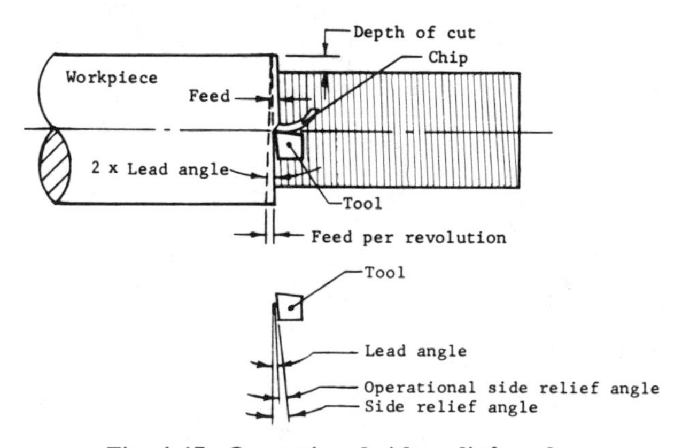

Fig. 4.45. Operational side relief angle.

side relief angle on a turning tool. An examination of Fig. 4.45 reveals that the lead angle also has an effect upon the so-called operational side relief angle. For this application, the operational side relief angle can be defined as the tool side relief angle minus the lead angle. This turns out to be the actual clearance angle between the tool and the workpiece during the cutting process.

Figure 4.46 shows the right triangle from which the lead angle can be extracted. The base of the triangle represents the lead, that is, the advance that the tool makes for each revolution of the workpiece. The lead is equivalent to the feed per revolution for a turning operation. The height of the triangle represents the circumference of the workpiece. The lead angle can be expressed as

$$\tan(\text{lead angle}) = \frac{\text{Lead}}{\text{Circumference}}$$

where lead = feed/rev = f, circumference = $\pi \times D$, and lead angle = L_a. As a result,

$$L_a = \tan^{-1}\left(\frac{f}{\pi \times D}\right) \tag{4.78}$$

The operational side relief angle can now be written as

$$O_a = S_a - L_a$$

where O_a is the operational relief angle, S_a is the side relief angle, and L_a is the lead angle, or

$$O_a = S_a - \tan^{-1}\left(\frac{f}{\pi \times D}\right) \tag{4.79}$$

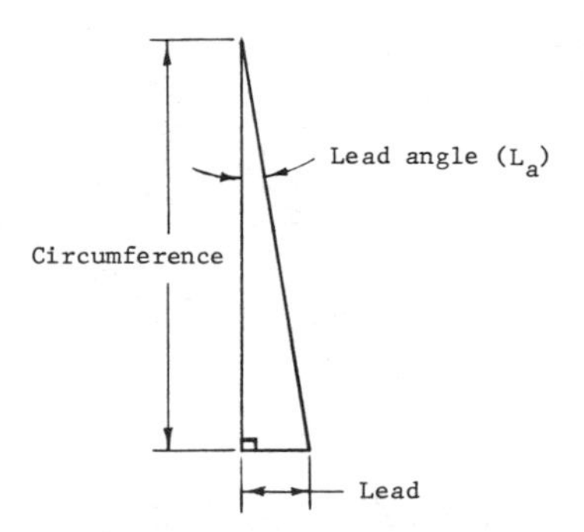

Fig. 4.46. Geometric model of lead angle.

The following example serves to illustrate how the lead angle reduces a side relief angle to an operational side relief angle.

EXAMPLE 4.35. A turning tool has ground into it a side relief angle of 6°. It is desired to determine the operational side relief angle for the machining of a 0.250-in. (6.35-mm) steel rod with a feed setting of 0.040 in./rev (1.016 mm/rev).

Substituting into Eq. 4.79 yields

$$\begin{aligned} O_a &= 6° - \tan^{-1}\left(\frac{0.040}{3.14 \times 0.250}\right) \\ &= 6° - 2.92° \\ &= 3.08° \end{aligned}$$

Of special interest in this case is the determination of the feed setting that reduces the operational side relief angle to zero. For this situation to exist, O_a must be equal to 0°. Under this circumstance, Eq. 4.79 can be written as

$$S_a = \tan^{-1}\left(\frac{f}{\pi \times D}\right)$$

from which

$$\tan S_a = \frac{f}{\pi \times D}$$

Isolating the feed yields

$$f = \pi \times D \times (\tan S_a) \tag{4.80}$$

EXAMPLE 4.36. What is the value of the feed that will reduce the operational side relief angle to zero for the data in example 4.35?

Substituting into Eq. 4.80 yields

$$\begin{aligned} f &= 3.14 \times 0.25 \times \tan 6° \\ &= 0.0825 \text{ in./rev } (2.1 \text{ mm/rev}) \end{aligned}$$

As can be seen in example 4.36, any feed setting that is above 0.0825 in./rev (2.1 mm/rev) will result in a negative operational side relief angle, causing contact on the lower portion of the tool flank with no contact at the cutting edge. This means that rather than cutting with contact at the cutting edge, the turning tool will contact the workpiece at the lower portion of the tool flank, causing a rubbing action during the operation.

The diameter of the workpiece has a major influence on the operational side relief angle. This can be seen by examining Eq. 4.79. A larger diameter with the same feed reduces the lead angle. Using the same feed and side relief angle as in example 4.35, but increasing the diameter from 0.25 in. (6.35 mm) to 2.00 in. (50.8 mm), reduces the lead angle to 0.364°. With the larger-diameter workpiece, the effect of the lead angle on reducing the operational side relief angle is minimal (0.364°), whereas with the smaller diameter it is significant, a 2.92° reduction.

The positioning of a turning tool relative to the center of a workpiece can have an influence on what may be considered the operational end relief angle. Figure 4.47 illustrates three examples of tool positioning relative to the horizontal centerline of the workpiece. In Fig. 4.47(a) the tool is set on center and the offset angle (ψ) is equal to zero. In this case, the operational end relief angle is the angle that has been ground into the tool. In Fig. 4.47(b), the offset angle (ψ) is set above the horizontal centerline, thus having an influence on reducing the end relief angle. Figure 4.47(c) shows the effect of setting the offset angle (ψ) below the horizontal centerline. In this case, the operational end relief angle is increased by the magnitude of the offset angle (ψ).

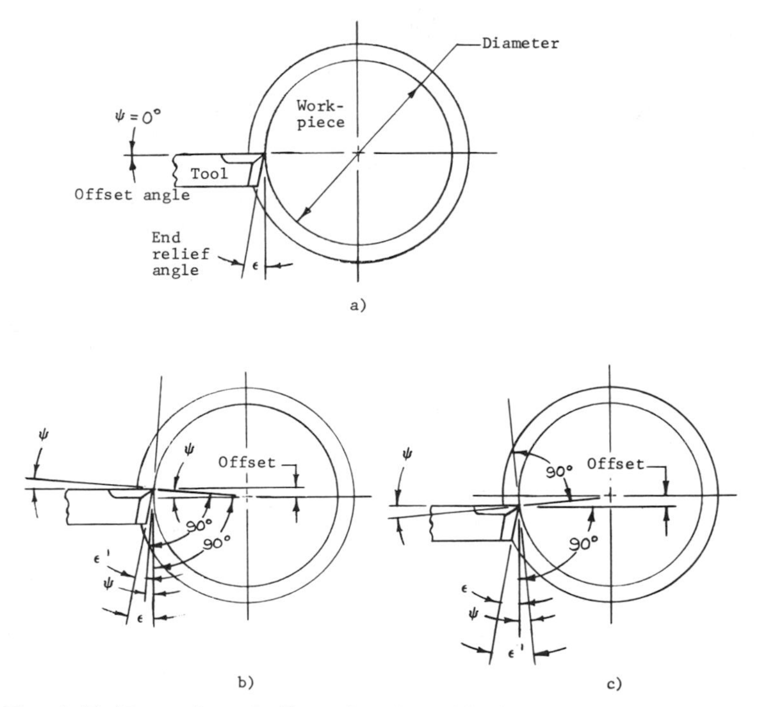

Fig. 4.47. Examples of effect of tool positioning on end relief angle.

Noting in Fig. 4.47 that the diameter represents the hypotenuse of the right triangle, the offset can be written as

$$\text{Off} = D \times \sin \psi \tag{4.81}$$

where Off is the vertical distance above or below the horizontal centerline, D is the diameter, and ψ is the offset angle. From Eq. 4.80, the offset angle can be isolated and expressed as

$$\psi = \sin^{-1}\left(\frac{\text{Off}}{D}\right) \tag{4.82}$$

An examination of Fig. 4.47(b) reveals that if the offset is above the horizontal centerline, then

$$\varepsilon' = \varepsilon - \psi \tag{4.83}$$

where ε' is the operational end relief angle, ε is the end relief angle, and ψ is the offset angle. For an offset below the horizontal centerline, the operational end relief angle is equal to

$$\varepsilon' = \varepsilon + \psi \tag{4.84}$$

To illustrate the importance of the offset position on the operational end relief angle, the following example is given.

EXAMPLE 4.37. It is desired to find the influence on the end relief angle of a 0.0625-in. (1.588-mm) turning tool offset, both above and below the horizontal centerline. The end relief angle ground into the tool is 7° and the diameter of the workpiece is 1 in. (25.4 mm).

The offset angle can be evaluated from Eq. 4.82 as

$$\begin{aligned}\psi &= \sin^{-1}\left(\frac{\text{Off}}{D}\right) = \sin^{-1}\left(\frac{0.0625}{1.000}\right)\\ &= 3.58^\circ\end{aligned}$$

With the offset above the horizontal centerline, the operational end relief angle is equal to

$$\begin{aligned}\varepsilon' &= \varepsilon - \psi\\ &= 7^\circ - 3.58^\circ = 3.42^\circ\end{aligned}$$

With the offset below the horizontal centerline, the operational end relief angle is equal to

$$\begin{aligned}\varepsilon' &= \varepsilon + \psi = 7^\circ + 3.58^\circ\\ &= 10.58^\circ\end{aligned}$$

As can be seen from the results of example 4.37, the offset has a significant influence on the operational end relief angle. Of interest also is that the offset influences the operational back rake angle. Figure 2.1 indicates tool angle designations. On a relative scale, as indicated in Fig. 4.47, the back rake angle is increased by the offset angle (ψ) if the offset (Off) is above the horizontal centerline and is decreased by the offset angle (ψ) if the offset (Off) is below the horizontal centerline.

There may be situations where the tool is set in a particular orientation, only to be offset by a deflection caused by the loading of the metal cutting process. A boring operation is a case of this type. Figure 4.48 illustrates the tool deflection on the boring bar caused by the cutting force. The deflection of the tool is designated as ΔY and is a response to the application of the cutting force ($\mathbf{F}_c$). If the tool was set at the horizontal centerline, then the deflection ΔY would be synonymous with the offset as discussed in example 4.37. In situations where the cutting force is high and the boring bar is slender, the deflection can be significant enough to appreciably alter the position of the tool. A means of evaluating the boring-bar deflection due to the action of the cutting force is through the following expression:

$$Y = \frac{\mathbf{F}_c \times L^3}{3 \times E \times I} \tag{4.85}$$

where $\mathbf{F}_c$ is the cutting force; E is the modulus of elasticity, which equals 30×10^6 psi (206.8×10^9 N/m^2) for steel; and I is the moment of inertia of cross-section area ($I = \pi \times D^4/64$ for a circular area). The following example illustrates the application of Eq. 4.85.

EXAMPLE 4.38. It is desired to determine the tool deflection on a boring bar caused by a cutting force loading. The steel boring bar has a 0.500-in. (12.7-mm) diameter and a length of 6 in. (152.4 mm). The cutting force has a value of 100 lb.

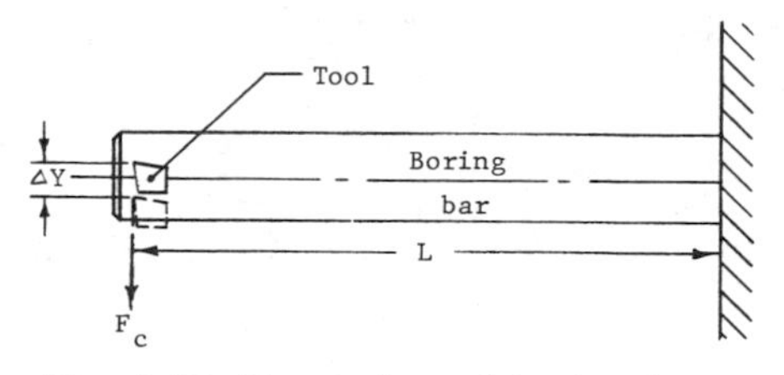

Fig. 4.48. Front view of boring bar.

Substituting into Eq. 4.85 yields

$$Y = \frac{F \times L^3}{3 \times E \times I} = \frac{100 \times 6^3 \times 64}{3 \times 30 \times 10^6 \times 3.14 \times (0.5)^4}$$

$$= 0.07827 \text{ in. } (1.988 \text{ mm})$$

As can be seen from the results of example 4.38, the deflection of a boring bar under a load can be significant.

4.14 THREE-DIMENSIONAL ANALYSIS OF RESULTANT CUTTING FORCE

Figure 4.49 illustrates the resolution of the resultant force that the tool exerts on the workpiece into three rectangular components, each parallel with its corresponding coordinate axis. The x component, labeled F_x, is the feed force and is a measure of the resistance the tool encounters during the cutting process in the direction of the feed. The y component, labeled F_y, is the cutting force and is a measure of the resistance the tool encounters in the direction of the cutting velocity. Lastly, the z component, labeled F_z, is the radial force and is a measure of the resistance the tool encounters in the direction of the radius of the workpiece.

Figure 4.50 shows the reactionary component forces to the forces that the tool exerts on the workpiece. Newton's third law of mechanics states that when two bodies exert forces on each other, these forces are equal in magnitude, opposite in direction, and act on the same line

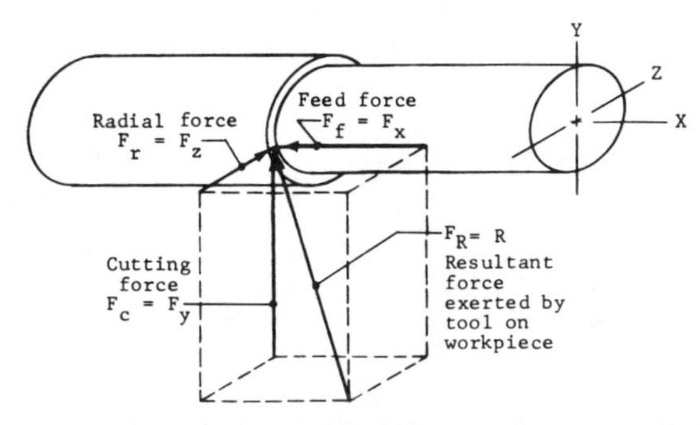

Fig. 4.49. Rectangular components of tool force acting on workpiece in turning operation.

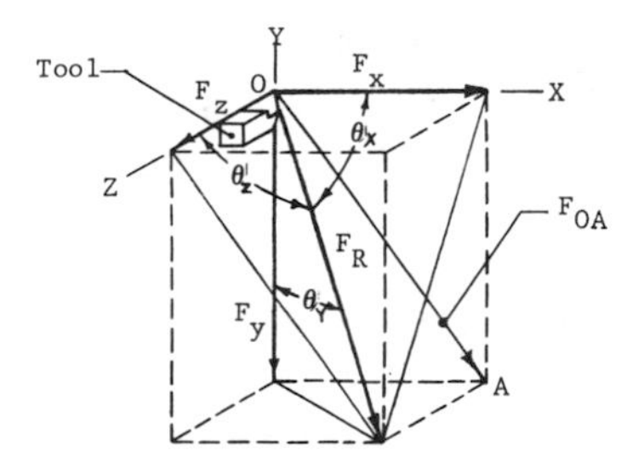

Fig. 4.50. Rectangular components of the workpiece forces acting on the tool.

of action. Figure 4.50 illustrates the component forces that the workpiece exerts on the cutting tool.

Of interest are the relationships of these component forces with the resultant cutting force. The magnitude of the rectangular components can be written as

$$F_x = F_R \cos\theta_x; \qquad F_y = F_R \cos\theta_y; \qquad F_z = F_R \cos\theta_z \tag{4.86}$$

By taking the ratio $1/F_R$ from each of the above expressions it can be written that

$$\frac{1}{F_R} = \frac{\cos\theta_x}{F_x} = \frac{\cos\theta_y}{F_y} = \frac{\cos\theta_z}{F_z} \tag{4.87}$$

The vectorial sum of F_x and F_y can be represented by a vector extending from O to A, as shown in Fig. 4.50. Its magnitude can be represented by applying the Pythagorean Theorem:

$$(\mathbf{F}_{OA})^2 = (F_x)^2 + (F_y)^2 \tag{4.88}$$

In turn, the resultant vectorial sum of $\mathbf{F}_{OA}$ and F_z must obviously be equal to F_R. Again, by applying the Pythagorean Theorem, it can be written that

$$(F_R)^2 = (\mathbf{F}_{OA})^2 + (F_z)^2$$

Substituting for $(\mathbf{F}_{OA})^2$ from Eq. 4.88 yields

$$(F_R)^2 = (F_x)^2 + (F_y)^2 + (F_z)^2$$

or

$$F_R = \sqrt{(F_x)^2 + (F_y)^2 + (F_z)^2} \tag{4.89}$$

By substituting values from Fig. 4.49, Eq. 4.89 can be written as

$$\mathbf{R} = \sqrt{(\mathbf{F}_f)^2 + (\mathbf{F}_c)^2 + (\mathbf{F}_r)^2} \tag{4.90}$$

By substituting in Eq. 4.89 the expressions for the component forces as listed in Eq. 4.87, the following identity emerges:

$$F_R = \sqrt{(F_R)^2 \cos^2 \theta_x + (F_R)^2 \cos^2 \theta_y + (F_R)^2 \cos^2 \theta_z}$$

or

$$1 = \cos^2 \theta_x + \cos^2 \theta_y + \cos^2 \theta_z \tag{4.91}$$

The three angles θ_x, θ_y, and θ_z define the direction of the resultant cutting force F_R. The cosines of these angles are known as the *directional cosines of the cutting force*. The following example demonstrates the application of the three-dimensional resolution of the resultant cutting force.

EXAMPLE 4.39. Determine the magnitude of the resultant force (F_R) and the angular displacements of the resultant from the three space axes (x, y, z). Dynamometer measurements record values of the component forces as: $\mathbf{F}_f = 210$ lb (934 N); $\mathbf{F}_c = 320$ lb (1423 N); and $\mathbf{F}_r = 50$ lb (222 N).

For this application, substitution into Eq. 4.90 yields

$$\mathbf{R} = \sqrt{(\mathbf{F}_f)^2 + (\mathbf{F}_c)^2 + (\mathbf{F}_r)^2} = \sqrt{(320)^2 + (210)^2 + (50)^2}$$
$$= 386 \text{ lb } (1717 \text{ N})$$

Angular displacements can be evaluated from Eq. 4.86 as

$$\theta_x = \cos^{-1}\left(\frac{\mathbf{F}_f}{F_R}\right) = \cos^{-1}\left(\frac{210}{386}\right)$$
$$= 57^\circ$$

$$\theta_y = \cos^{-1}\left(\frac{\mathbf{F}_c}{F_R}\right) = \cos^{-1}\left(\frac{320}{386}\right)$$
$$= 85.2^\circ$$

$$\theta_z = \cos^{-1}\left(\frac{\mathbf{F}_r}{F_R}\right) = \cos^{-1}\left(\frac{50}{386}\right)$$
$$= 82.6^\circ$$

Experiments have revealed that the cutting force $\mathbf{F}_c$, the feed force $\mathbf{F}_f$, and the radial force $\mathbf{F}_r$ are functions of the feed and depth of cut and can be expressed in the form

$$\text{Force} = \text{Constant} \times \text{Feed}^{(\text{exponent}_f)} \times \text{Depth of cut}^{(\text{exponent}_d)}$$

where the constant, the exponent$_f$, and the exponent$_d$ are fixed for a tool of a specified shape, but vary to different values as the shape of the tool changes[3]. In other words, it can be written that for a tool of a particular shape,

$$\mathbf{F}_c = C_c \times f^{n_1} \times d^{n_2} \quad (4.92)$$

$$\mathbf{F}_f = C_f \times f^{n_3} \times d^{n_4} \quad (4.93)$$

and

$$\mathbf{F}_r = C_r \times f^{n_5} \times d^{n_6} \quad (4.94)$$

where $\mathbf{F}_c$ is the cutting force, $\mathbf{F}_f$ is the feed force, $\mathbf{F}_r$ is the radial force, C_c is a constant for cutting force, C_f is a constant for feed force, C_r is a constant for a radial force, n_1 is the exponent of feed for cutting force, n_2 is the exponent of depth of cut for cutting force, n_3 is the exponent of feed for feed force, n_4 is the exponent of depth of cut for feed force, n_5 is the exponent of feed for radial force, and n_6 is the exponent of depth of cut for radial force. Boston and Kraus found that for a turning operation on annealed low-carbon steel using a high-speed-steel tool with a signature (designation—see Fig. 2.1) of 8-14-6-6-6-0-3-3/16 R, running at approximately 80 ft/min (24.38 m/min), the equations for cutting force, feed force, and radial force were of the form

$$\mathbf{F}_c = 58{,}000 \times f^{0.68} \times d^{0.83} \quad (4.95)$$

$$\mathbf{F}_f = 31{,}800 \times f^{0.57} \times d^{1.31} \quad (4.96)$$

$$\mathbf{F}_r = 7250 \times f^{0.68} \times d^{0.47} \quad (4.97)$$

The determination of the resultant force from the individual component forces is demonstrated by the following example.

EXAMPLE 4.40. A turning operation running at 80 ft/min (24.38 m/min), with a feed of 0.015 in. (0.381 mm) and a depth of cut of 0.250 in. (6.35 mm), has equations for the component forces in the form of Eqs. 4.95–4.97. From these data, determine:

(a) The cutting force.
(b) The feed force.
(c) The radial force.
(d) The resultant force.
(e) The horsepower required of the cut.
(f) The unit horsepower.
(g) The motor horsepower required if the machine has an efficiency of 85% and a tare horsepower of 0.75.

[3] O. W. Boston and C. E. Kraus, "A Study of the Turning of Steel Employing a New-Type Three-Component Dynamometer," *Trans. ASME*, RP-58-1, 1936, pp. 47–53.

Substituting into Eq. 4.95 yields

$$\mathbf{F}_c = 58{,}000 \times (0.015)^{0.68} \times (0.25)^{0.83}$$
$$= 1053 \text{ lb } (4684 \text{ N})$$

From Eq. 4.96,

$$\mathbf{F}_f = 31{,}800 \times (0.015)^{0.57} \times (0.25)^{1.31}$$
$$= 472 \text{ lb } (2{,}100 \text{ N})$$

From Eq. 4.97,

$$\mathbf{F}_r = 7250 \times (0.015)^{0.68} \times (0.25)^{0.47}$$
$$= 217 \text{ lb } (966 \text{ N})$$

The resultant force can be determined by substituting into Eq. 4.90:

$$\mathbf{R} = \sqrt{(\mathbf{F}_f)^2 + (\mathbf{F}_c)^2 + (\mathbf{F}_r)^2}$$
$$= \sqrt{(472)^2 + (1052)^2 + (217)^2}$$
$$= 1173 \text{ lb } (5218 \text{ N})$$

From Eq. 4.69, the horsepower required of the cut is

$$\text{hp}_c = \frac{\mathbf{F}_c \cdot \mathbf{V}_c}{33{,}000} = \frac{1053 \times 80}{33{,}000}$$
$$= 2.55 \text{ horsepower } (1.902 \text{ kW})$$

Substituting into Eq. 4.72 yields the unit horsepower as being equal to

$$U_p = \frac{\text{hp}_c}{R_v} = \frac{2.55}{3.6}$$
$$= 0.708 \text{ hp/in.}^3\text{/min } (1.934 \text{ kW/cm}^3\text{/sec})$$

where

$$R_v = 12 \times f \times d \times \mathbf{V}_c$$
$$= 12 \times 0.015 \times 0.25 \times 80$$
$$= 3.6 \text{ in.}^3\text{/min } (0.9828 \text{ cm}^3\text{/sec})$$

Finally, from Eq. 4.77,

$$\text{hp}_m = \frac{\text{hp}_c}{E_{\text{ff}}} + \text{hp}_t = \frac{2.55}{0.85} + 0.75$$
$$= 3.75 \text{ horsepower } (2.8 \text{ kW})$$

Since the constants (c) and exponents (n) of the component force equations vary with the material being cut as well as with the shape of the tool, specifications for a particular operation needs to involve an experimental technique. The feed (f) and the depth of cut (d) are independent variables, that is, their values do not depend on each other. Therefore, by holding one of the independent variables constant, we can evaluate the effect of changes of the other independent variable on the component force. In other words, by holding the depth of cut constant, the influence of the feed on the cutting force can be expressed as

$$\mathbf{F}_c = C_c \times f^{n_1} \times d^{n_2}$$

or

$$\mathbf{F}_c = C_1 \times f^{n_1} \tag{4.98}$$

where

$$C_1 = C_c \times d^{n_2}$$

If experimental data are compiled between the cutting force and the feed, with the depth of cut fixed, then the constant (C_1) and the exponent (n_1) of the feed can be evaluated. To determine the exponent (n_2) of the depth of cut, a similar technique may be employed. This can be accomplished by keeping the feed constant and compiling experimental data between the cutting force and the depth of cut. In this case, the cutting force equation changes from

$$\mathbf{F}_c = C_c \times f^{n_1} \times d^{n_2}$$

to

$$\mathbf{F}_c = C_2 \times d^{n_2} \tag{4.99}$$

where

$$C_2 = C_c \times f^{n_1}$$

Assume that in an effort to determine the relationship of the cutting force as a function of the feed and depth of cut, the following experiment was conducted. Dynamometer readings of the cutting force were taken with the depth of cut fixed and with four different settings of the feed. Then additional dynamometer readings were taken, but this time with the feed fixed and with four different settings of the depth of cut. The experimental data are listed in Table 4.6. The value of the constant C_1 and the exponent n_1 in Eq. 4.98 can be evaluated by substituting experimental data from Table 4.6. From tests 1 and 2, it can be written that

$$\mathbf{F}_c = C_1 \times f^{n_1}$$
$$125.7 = C_1 \times (0.005)^{n_1}$$

Table 4.6. Cutting Force Experimental Data

Test	Feed		Depth of Cut		Cutting Force	
	(in./rev)	(mm/rev)	(in.)	(mm)	(lb)	(N)
1	0.005	0.127	0.050	1.27	125.7	559
2	0.010	0.254	0.050	1.27	218.8	973
3	0.015	0.381	0.050	1.27	302.3	1344
4	0.020	0.508	0.050	1.27	380.7	1693
5	0.010	0.254	0.050	1.27	218.8	973
6	0.010	0.254	0.100	2.54	422.8	1881
7	0.010	0.254	0.150	3.81	621.4	2764
8	0.010	0.254	0.200	5.08	816.7	3633

and

$$218.8 = C_1 \times (0.010)^{n_1}$$

Isolating C_1 yields

$$C_1 = \frac{125.7}{(0.005)^{n_1}} = \frac{218.8}{(0.010)^{n_1}} \tag{1}$$

or

$$\left(\frac{0.010}{0.005}\right)^{n_1} = \frac{218.8}{125.7} = 1.7407$$

Solving for n_1 yields

$$n_1 \log 2 = \log 1.7407$$

$$n_1 = \frac{\log 1.7407}{\log 2} = \frac{0.2407}{0.301} = 0.8$$

By substituting in Eq. 1 for the value of n_1, C_1 is found to be equal to

$$C_1 = \frac{125.7}{(0.005)^{0.8}} = \frac{125.7}{0.0144} = 8712.8$$

For the cutting operation described in Table 4.6, Eq. 4.98 can now be expressed as

$$F_c = 8712.8 \times f^{0.8}$$

The constant and exponent in Eq. 4.99 can also be evaluated by using data from Table 4.6. Taking the values from tests 6 and 7 yields

$$\mathbf{F}_c = C_2 \times d^{n_2}$$

$$422.8 = C_2 \times (0.100)^{n_2}$$

$$621.4 = C_2 \times (0.150)^{n_2}$$

from which

$$C_2 = \frac{422.8}{(0.100)^{n_2}} = \frac{621.4}{(0.150)^{n_2}} \tag{2}$$

Isolating n_2, we obtain

$$\left(\frac{0.150}{0.100}\right)^{n_2} = \frac{621.4}{422.8}$$

$$(1.5)^{n_2} = 1.4697$$

$$n_2 = \frac{\log 1.4697}{\log 1.5} = \frac{0.1672}{0.17609} = 0.95$$

C_2 can now be evaluated from Eq. 2 as

$$C_2 = \frac{422.8}{(0.100)^{0.95}} = \frac{422.8}{0.1122} = 3768.2$$

For the cutting operation expressed by the data in Table 4.6, Eq. 4.99 can be written as

$$\mathbf{F}_c = 3768.2 \times d^{0.95}$$

With the availability of the value of the exponents in Eq. 4.92, the numerical value of the cutting constant can easily be determined for this particular cutting operation by substituting data from Table 4.6. We can illustrate by using data from test 8:

$$\mathbf{F}_c = C_c \times f^{n_1} \times d^{n_2}$$

$$816 = C_c \times (0.010)^{0.8} \times (0.200)^{0.95}$$

$$C_c = \frac{816.7}{0.02512 \times 0.21676} = 150{,}000$$

As a result, the cutting force for the operation described by the data of Table 4.6 can now be written as

$$\mathbf{F}_c = 150{,}000 \times f^{0.8} \times d^{0.95}$$

A point of interest is that the numerical values of the exponents n_1 and n_2 can also be determined graphically by plotting the experimental data. To illustrate, Eq. 4.98 can be written as

$$\mathbf{F}_c = C_1 \times f^{n_1}$$

or for convenience as

$$\log \mathbf{F}_c = n_1 \log f + \log C_1 \tag{4.100}$$

which has the form of the equation of a straight line

$$y = mx + b$$

When plotted on log–log graph paper, Eq. 4.98 appears as a straight line, where the exponent n_1 represents the slope of the line. Figure 4.51 represents the plot. As can be seen,

$$\text{Slope} = n_1 = \tan\theta_1 = \frac{\log \mathbf{F}_{c4} - \log \mathbf{F}_{c1}}{\log f_4 - \log f_1} \tag{4.101}$$

Solving for the value of the exponent n_1 from the example listed in Table 4.6 by using data from tests 1 and 4 gives

$$n_1 = \frac{\log 380.7 - \log 125.7}{\log 0.020 - \log 0.005}$$

$$= \frac{2.5806 - 2.0993}{-1.699 - (-2.301)} = \frac{0.4813}{0.602}$$

$$= 0.8$$

This value could also be attained through a direct angular measurement from Fig. 4.51, where

$$n_1 = \tan\theta_1 = \tan 38.66^\circ = 0.8$$

With a similar graphical approach, the value of the exponent n_2 can also be found. The experimental values represented by Eq. 4.99, plotted on log–log graph paper, appear in the form of a straight line.

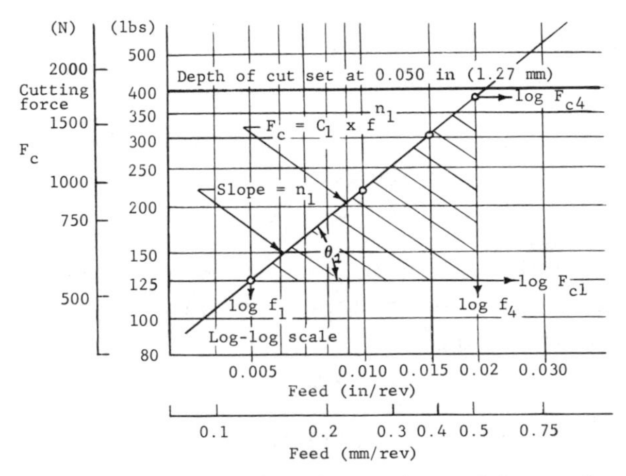

Fig. 4.51. Graphical representation of experimental data on log–log scale.

This is shown in Fig. 4.52. As can be seen,

$$\text{Slope} = n_2 = \tan\theta_2 = \frac{\log \mathbf{F}_{c8} - \log \mathbf{F}_{c5}}{\log d_8 - \log d_5} \tag{4.102}$$

Solving for the exponent n_2 by using data form tests 5 and 8 gives

$$n_2 = \frac{\log 816.7 - \log 218.8}{\log 0.2 - \log 0.05} = \frac{2.912 - 2.340}{-0.699 - (-1.301)}$$

$$= \frac{0.572}{0.602} = 0.95$$

The value of n_2 can also be extracted by a direct angular measurement from the graph. As a result,

$$n_2 = \tan\theta_2 = \tan 43.53^\circ = 0.95$$

With the availability of the exponents n_1 and n_2, the value of the constant C_c in Eq. 4.92 can easily be evaluated by taking any set of values from the experimental data. As an example, using data from test 1 yields

$$\mathbf{F}_c = C_c \times f^{n_1} \times d^{n_2}$$

$$125.7 = C_c \times (0.005)^{0.8} \times (0.050)^{0.95}$$

$$C_c = \frac{125.7}{0.014427 \times 0.05808} = 150{,}000$$

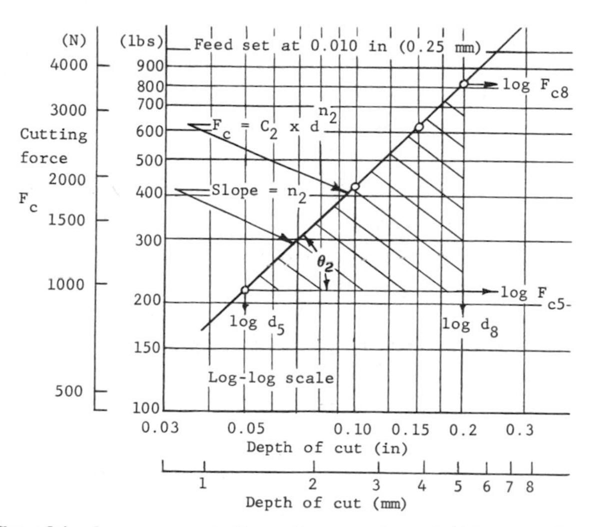

Fig. 4.52. Graphical representation of experimental data on log–log scale.

It is then confirmed that for the given experimental data, the cutting force equation can be written as

$$\mathbf{F}_c = 150{,}000 \times f^{0.8} \times d^{0.95}$$

The values of the constants and exponents for the feed force (Eq. 4.93) and the radial force (Eq. 4.94) can be determined from experimental data in a fashion similar to that employed for the cutting force (Eq. 4.92).

4.15 SUMMARY OF THE MECHANICS OF THE CUTTING PROCESS

The analysis and application of the forces acting between the work material and the tool constitutes the mechanics of the cutting process. Dynamometers, which are force-sensing instruments, provide a means of measuring cutting forces. Through a convenient design of a dynamometer, the resultant cutting force can be resolved into three space components. These components, in turn, enable the resolution of additional directional components that provide an insight into the interaction of the tool and workpiece.

The tool–chip interface force is a case in point. By determining the magnitude of the force acting parallel to the face of the tool, a measure of the tool–surface resistance to the sliding motion of the chip can be attained. This, in turn, enables us to determine the coefficient of friction, which reflects directly on the interaction of the tool and work-piece. In this way, a measure of the quality of the tool surface is determined.

Free-body diagrams of the tool, chip, and workpiece show how the resultant cutting force is transmitted during the cutting process. The free-body diagrams also provide a means of isolating component forces in directions covenient for analysis. Force magnitudes can also be extracted from free-body diagrams. A case in point is the evaluation of the magnitude of the force that acts in the direction of the chip shear plane. This force, distributed over the area of the shear plane, provides a measure of the plastic flow shear stress state that exists during the metal cutting operation.

Vector polygon graphical layouts of the component forces provide not only a visual display of the forces but also a means of expressing forces in vector summation equations. The graphical layouts are justified by the first and third laws of Newton's equilibrium state.

This analysis leads to the determination of the direction and magnitude of individual component forces that are represented in vector form.

The resolution of the resultant cutting force into three sets of perpendicular components extends the analysis of evaluating the effects of a component force in a particular direction. Force action along the shear plane, along the face of the tool, and along the cutting direction leads to quantification of cutting characteristics.

As an example, force action on the shear plane is a reflection of the shear stress of the work material and also influences the cutting force that acts on a plane perpendicular to the direction of the cutting force. The plane over which the cutting force is distributed is equal to the feed multiplied by the depth of cut, and the ratio of the cutting force divided by this area is called the *chip pressure*. It has been shown that the shear angle is a function of the tool side-rake angle and that it has a major influence on the chip pressure.

Force action on the face of the tool can be interpreted as a measure of the quality of the surface of the tool. To illustrate, a tool that provides a low friction force while the chip slides over it can be expected to use less energy and thus generate lower temperatures in the cutting process. The term *coefficient of friction* is a measure of the ratio of the frictional force to the force acting normal to the frictional surface. This term does not have a fixed value when considering a specific tool machining operation with a particular material. Rather, it has been shown experimentally that the coefficient of friction varies as a function of the cutting velocity, which, in turn, is related to the temperature acting on the tool and the workpiece.

Force action in the direction of the cutting velocity is of importance because it is related to the energy expended during the metal cutting operation. The product of the cutting force multiplied by the cutting velocity is expressed in units of work per unit time, which is a measure of power.

Velocity components other than the cutting velocity are also worth of consideration. As an example, the chip velocity is related to the cutting velocity through the shear angle and is affected by the chip thickness. Obviously, the relationship is inverse, that is, a thicker chip will produce a lower chip velocity.

An energy analysis of the major components of the resultant force, namely, the cutting force ($\mathbf{F}_c$), the feed force ($\mathbf{F}_f$), and the radial force ($\mathbf{F}_r$) reveals that the predominant amount of energy used in a metal cutting operation is consumed by the cutting force. As a result, for practical purposes of analysis, energy absorption by the feed and radial forces are negligible and need not be considered.

Rotary tools, such as drills and milling cutters, have similar component force designations as those for a single-point tool. The reason for this is that multiple-point tools can be reduced in the analysis of force components to a series of single-point tools spaced along the circumference of the tool. In analyzing rotary tools, the work performed is equal to the product of the torque and the angular displacement through which the torque acts. This is an important point insofar as confusion oftentimes develops with respect to the balance of the units. In this case, the radian is considered unitless and the work has assigned to it the same units as the torque.

The amount of work performed per unit time is the definition of *power*. It is the rate at which work (energy) is performed. The power per unit volume required for a metal cutting operation is the definition of *unit horse power*. This value is a measure of the power required to machine at a volumetric rate of unity. It has been found that the unit horsepower is related to the hardness of the work material. A harder material requires a higher unit horsepower during the metal cutting operation.

The amount of power required at the cut is not equal to the total power required to perform an operation on a machine tool. This is because (1) all machine tools require a certain amount of power to run in the idling mode and (2) all machine tools are less than 100% efficient. Therefore, in calculating the capacity of a machine in terms of motor horsepower, the machine efficiency as well as the tare (idling) horsepower must be taken into consideration.

Another point of interest is the effect that the feed and positioning of a tool have on the clearance angles that are ground into the tool. What may seem to be a desirable side relief angle in a turning operation can be eliminated completely if the lead angle of the feed is greater than the clearance angle. The result is a rubbing action rather than one of cutting. A similar situation can develop when the tool is set too high above center. This condition can eliminate the end relief angle in a turning operation. Deflections of tools, such as boring bars, under the forces of cutting can also influence the clearance angles. This takes place when the tool moves from its unloaded position as a result of absorbing the forces acting on it.

The relationship of the components of the resultant cutting force to cutting conditions, such as the feed and the depth of cut, can be extracted from experimental data. As an example, if one compiles data on the cutting force as a function of the feed, with the depth of cut fixed, then the relationship between cutting force and feed can be established. In a similar fashion, the relationship between cutting force and depth of cut can be established. This can be done if data are compiled where the cutting force is measured as a function of the depth of cut, with the feed fixed. By combining the test results, expressions can be written relating the component cutting forces with cutting conditions.

PROBLEMS

PROBLEM 4.1. For a two-dimensional cutting force analysis similar to that shown in Fig. 4.3, determine the magnitude and direction of the resultant cutting force if dynamometer measurements are 1690 N for the cutting force ($\mathbf{F}_c$) and 556 N for the feed force ($\mathbf{F}_f$).

Answers: R = 1779 N
$\theta = 18.2°$

PROBLEM 4.2. Dynamometer readings register values of 450 lb for the cutting force ($\mathbf{F}_c$) and 156 lb for the feed force ($\mathbf{F}_f$). The cutting tool is found to have a rake angle of 10°. From this information, determine:

(a) The resultant force.
(b) The angle (θ) at which the resultant force acts.
(c) The friction force (**F**).
(d) The normal force (**N**).
(e) The coefficient of friction (μ).

Answers: **R** = 476.3 lb
$\theta = 19.12°$
F = 231.6 lb
N = 4.5.8 lb
$\mu = 0.557$

PROBLEM 4.3. A turning operation registers a cutting force of 1112 N and a feed force of 333.6 N. It is also found that the chip thickness is 1.143 mm for a corresponding feed of 0.381 mm/rev. The tool performing the cutting has a rake angle of 10°. From this information determine:

(a) The resultant force (**R**).
(b) The angle at which the resultant force acts (θ).
(c) The shear angle (α).
(d) The shear force ($\mathbf{F}_s$).
(e) The force normal to the shear plane ($\mathbf{F}_n$).

Answers: R = 1161 N
$\theta = 16.7°$
$\alpha = 19.02°$
$\mathbf{F}_s = 942.6$ N
$\mathbf{F}_n = 677.8$ N

PROBLEM 4.4. For a metal cutting operation, the shear stress is to be evaluated by using dynamometer readings. It is found through chip thickness measurements that the shear angle is 20°. Other data compiled include: $\mathbf{F}_c = 800$ lb; $\mathbf{F}_f = 275$ lb; $f = 0.020$ in./rev; and $d = 0.375$ in.

Answer: $S_s = 29{,}993$ psi

PROBLEM 4.5. A metal cutting operation is in the process of being evaluated for the magnitude of the three sets of perpendicular force components shown in Fig. 4.12, into which the resultant force can be resolved. Dynamometer readings give values of 1780 N for the cutting force ($\mathbf{F}_c$) and 670 N for the feed force ($\mathbf{F}_f$). The chip thickness ratio (r_a) is found to be equal to 0.27 and the tool rake angle is 5°. With the availability of these data, calculate:

(a) The shear angle (α).
(b) The resultant force ($\mathbf{R}$).
(c) The angle of kinetic friction (ϕ).
(d) The frictional force ($\mathbf{F}$).
(e) The normal force ($\mathbf{N}$).
(f) The angle K as shown in Fig. 4.12.
(g) The shear force ($\mathbf{F}_s$).
(h) The force normal to the shear plane ($\mathbf{F}_n$).

Answers: $\alpha = 15.4°$
$\mathbf{R} = 1902$ N
$\phi = 25.63°$
$\mathbf{F} = 822.7$ N
$\mathbf{N} = 1714.9$ N
$K = 36.03°$
$\mathbf{F}_s = 1538.2$ N
$\mathbf{F}_n = 1118.8$ N

PROBLEM 4.6. Using the information provided in problem 4.4, calculate the chip pressure resistance and compare it with the shear stress.

Answers: $P_c = 106{,}666$ psi
$R_{sc} = 0.28$

PROBLEM 4.7. It is desired in a metal cutting operation to determine the value of the cutting force and the coefficient of friction by using the yield shear stress, which is found to be 220×10^6 Pa for the material being cut. For the operation, a 5°-side-rake-angle tool produces a shear angle of 20°. The feed is set at 0.457 mm/rev and the depth of cut is 6.35 mm. Assume that $\mathbf{F}_n/\mathbf{F}_s = 1$.

Answers: $\mathbf{F}_c = 2391$ N
$\mu = 0.577$

PROBLEM 4.8. Calculate the expected cutting force for an operation where the shear stress is 32,000 psi and the shear angle is 18°. The feed is set at 0.010 in./rev and the depth of cut is 0.100 in. Assume that $\mathbf{F}_n/\mathbf{F}_s = 1$.

Answer: $\mathbf{F}_c = 130.6$ lb

PROBLEM 4.9. Determine the coefficient of friction and the angle of kinetic friction for a metal cutting operation using a 10°-rake-angle tool. Dynamometer readings are 1500 N for the cutting force ($\mathbf{F}_c$) and 800 N for the feed force ($\mathbf{F}_f$).

Answers: $\mu = 0.783$
$\phi = 38.07°$

PROBLEM 4.10. Calculate the chip velocity and the shear velocity for an operation being machined with a tool that has a side rake angle of 5°. The cutting velocity is 200 ft/min. Other machine settings include: chip thickness equal to 0.032 in. and feed equal to 0.014 in.

Answers: $\mathbf{V}_p = 87.47$ ft/min
$\mathbf{V}_s = 211.2$ ft/min

PROBLEM 4.11. Determine the amount of work performed by each of the three components of the resultant force for one revolution of a turning operation. The component forces are:

$$\mathbf{F}_c = 1000 \text{ N}, \quad \mathbf{F}_f = 350 \text{ N}, \quad \text{and} \quad \mathbf{F}_r = 100 \text{ N}$$

The diameter of the workpiece is 60 mm and is being machined at 50 m/min, with a feed of 0.4 mm/rev and a depth of cut of 6.0 mm.

Answers: $W_c = 188.4$ joules
$W_f = 0.14$ joules
$W_r = 0$ joules

PROBLEM 4.12. Using a graphical technique, confirm the answers of problem 4.10.

Answers: $\mathbf{V}_p = 87.47$ ft/min
$\mathbf{V}_s = 211.2$ ft/min

PROBLEM 4.13. A drill has acting on it a torque of 20 N-m and a thrust of 4000 N. The feed is set at 0.25 mm/rev. For one revolution of the drill, calculate the work done by the torque and by the thrust force.

Answers: $W_t = 125.66$ joules
$W_{st} = 1.0$ joules

PROBLEM 4.14. Solve for the torque and the thrust of a standard $\frac{3}{4}$-in. drill that is machining 200-Bhn steel with a feed setting of 0.010 in./rev.

Answers: $T = 368$ in.-lb
$T_{st} = 1455.7$ lb

PROBLEM 4.15. It is desired to determine the amount of power necessary to perform a given cut. The cutting velocity is 150 m/min, the feed is 0.5 mm/rev, the depth of cut is 5.0 mm, and the cutting force is 5000 N.

Answer: $P = 12.5$ kW

PROBLEM 4.16. A 3-in.-diameter milling cutter with eight cutting edges is programmed to rotate at 120 rpm. Calculate the cutting velocity and the feed per tooth for a workpiece feed of 2.5 in./min.

Answers: $\mathbf{V}_c = 94.2$ ft/min
$f_{\mathbf{tooth}} = 0.0026$ in.

PROBLEM 4.17. A milling operation is set for a depth of cut of 12.7 mm, and a torque measurement indicates a value of 135.6 N-m. The rpm is 100 and the work feed is 44.45 mm/min. From these data, calculate the power required for the cutting operation.

Answer: $P = 1.42$ kW

PROBLEM 4.18. Determine the amount of horsepower required for a $\frac{3}{4}$-in.-diameter drilling operation that generates a torque of 310 in.-lb when running at 450 rpm.

Answer: hp = 2.21 horsepower

PROBLEM 4.19. A turning operation is observed to have a cutting force of 1500 N and a cutting velocity of 75 m/min. From these data, calculate the power required for the cutting operation.

Answer: $P = 1.875$ kw

PROBLEM 4.20. A 1-in.-diameter end mill is set to cut a 0.500-in.-deep by 1-in.-wide slot in a steel workpiece that has a Bhn of 180. The end mill is set to run at 80 ft/min with a feed of 0.0025 in./tooth. The end mill has six teeth (cutting edges). From these data, calculate:

(a) The feed set on the machine tool.
(b) The volumetric rate of metal removal.
(c) The horsepower consumed in the operation.
(d) The torque acting on the end mill by using a unit horsepower value from Table 4.5.
(e) The average cutting force acting on the end mill.

Answers: $f = 4.586$ in./min
$R_v = 2.293$ in.^{3}/min
$hp_c = 1.72$ horsepower
$T = 354.6$ in.-lb
$\mathbf{F}_c = 709.5$ lb

PROBLEM 4.21. A 12-mm-diameter high-speed standard drill is machining steel that possesses a unit power of 3.0 kW/cm^3/sec. The cutting velocity is set at 12 m/min and the feed rate is 40 mm/min. Neglecting the effects of the chisel edge, determine the power required for this operation as well as the torque acting on the drill.

Answers: $P = 0.264$ kW
$T = 7.93$ N-m

PROBLEM 4.22. Reference sources reveal that a medium carbon steel workpiece material has a unit horsepower of 0.8 hp/in.^{3}/min. For a turning operation, the cutting velocity is set at 500 ft/min with a corresponding feed of 0.022 in./rev and a depth of cut of 0.210 in. From the given data, calculate:

(a) The cutting force.
(b) The volumetric rate of machining.
(c) The horsepower consumed by the cut.

Answers: $\mathbf{F}_c = 1463$ lb
$R_v = 27.72$ in.^{3}/min
$hp_c = 22.176$ horsepower

PROBLEM 4.23. It is desired to determine the overall efficiency of a machine tool. A wattmeter is set up to measure the input power and a dynamometer is used to measure the cutting force. For a cutting velocity of 50 m/min, it is found that the wattmeter reading is 2 kW and the dynamometer reading is 2000 N.

Answer: $E_{ff} = 83.4\%$

PROBLEM 4.24. It is desired to determine the required motor horsepower of a turning operation. The machine being used has a power efficiency of 85% and a tare horsepower of 0.5. The cutting velocity is set at 200 ft/min, with a feed of 0.015 in./rev and a depth of cut of 0.250 in. The unit horsepower of the material being cut is found to be 1.2 hp/in.^{3}/min. With this information, determine:

(a) The volumetric rate of metal removal.
(b) The horsepower required at the cut.
(c) The motor horsepower.

Answers: $R_v = 9$ in.^{3}/min
$hp_c = 10.8$ horsepower
$hp_m = 14$ horsepower

PROBLEM 4.25. Determine the motor horsepower required for an operation where the required power to perform the cut is 5 kW, the machine efficiency is 80%, and the tare power is 0.3 kW.

Answer: $P = 6.55$ kW

PROBLEM 4.26. Determine the tool deflection on a steel boring bar that has a 0.75-in. diameter, a 5-in. length, and a 200-lb cutting force acting on it.

Answer: $Y = 0.018$ in.

PROBLEM 4.27. Dynamometer readings record the following values of component forces for a turning operation: $\mathbf{F}_f = 950$ N, $\mathbf{F}_c = 1500$ N, and $\mathbf{F}_r = 225$ N. From this information, calculate the resultant force ($\mathbf{R}$).

Answer: $\mathbf{R} = 1789.7$ N

PROBLEM 4.28. A turning operation running at 80 ft/min, with a feed of 0.020 in./rev and a depth of cut of 0.375 in., has equations for component forces in the form of Eqs. 4.95–4.97. From these data, determine:

(a) The cutting force.
(b) The feed force.
(c) The radial force.

Answers: $\mathbf{F}_c = 1796$ lb
$\mathbf{F}_f = 946$ lb
$\mathbf{F}_r = 320$ lb

PROBLEM 4.29. Using the experimental data represented in Table 4.6, write the equation for the cutting force using SI units, where the feed is given in millimeters per revolution, the depth of cut is given in millimeters, and the cutting force is given in newtons..

Answer: $\mathbf{F}_c = 2321 \times f^{0.8} \times d^{0.95}$

PROBLEM 4.30. A metal cutting operation is described by the answer given in Problem 4.29.

(a) Calculate the cutting force for a feed of 0.127 mm/rev and a depth of cut of 1.27 mm.
(b) If the feed is increased to 0.254 mm/rev and the depth of cut is increased to 5.08 mm, determine the corresponding cutting force.

Answers: (a) $\mathbf{F}_c = 559$ N
(b) $\mathbf{F}_c = 3633$ N

BIBLIOGRAPHY

Baumeister, Theodore, *Marks' Standard Handbook for Mechanical* Engineers, McGraw-Hill, New York, 1978.

Beer, Ferdinand P., and Johnston, E. Russell, Jr., *Mechanics for Engineers—Statics*, McGraw-Hill, New York, 1976.

Boston, O. W., and Kraus, C. E., "A Study of the Turning of Steel Employing a New-Type Three-Component Dynamometer," *Trans. ASME*, RP-58-1, 1936.

Boston, Orlan William, *Metal Processing*, John Wiley & Sons, New York, 1951.

Cook, Nathan H., *Manufacturing Analysis*, Addison-Wesley, Reading, Massachusetts, 1966.

Costa, Leslie F., Jr., "A Measurement of the Coefficient of Friction as a Function of Cutting Speed," Senior Design Project, Southeastern Massachusetts University, North Dartmouth, Massachusetts, 1977.

Dallas, Daniel B., *Tool and Manufacturing Engineers Handbook*, Society of Manufacturing Engineers, Dearborn, Michigan, 1976.

Hoffman, Edward G., *Fundamentals of Tool Design*, Society of Manufacturing Engineers, Dearborn, Michigan, 1984.

Machining Data Handbook, Metcut Research Associates, Cincinnati, Ohio, 1972.

Manual on Cutting Metal, The American Society of Mechanical Engineers, New York, 1952.

Merchant, M. Eugene, "Basic Mechanics of the Metal-Cutting Process," *J. Appl. Mech.*, Vol. 15, September 1944.

Merchant, M. Eugene, "Mechanics of the Metal Cutting Process II, Plasticity Conditions in Orthogonal Cutting," *J. Appl. Mech.*, Vol. 16, June 1945.

Oberg, E., Jones, F. D., and Horton, H., *Machinery's Handbook*, 22nd ed. Industrial Press, New York, 1984.

Sen, Gopal Chandra, and Bhattacharyya, Amitabha, *Principles of Metal Cutting*, New Central Book Agency, Calcutta, India, 1969.

Shaw, M. C., and Oxford, C. J., "On the Drilling of Metals 2—The Torque and Thrust in Drilling," *Trans. ASME*, Vol. 79, 1957.

Shaw, Milton C., *Metal Cutting Principles*, MIT Press, Cambridge, Massachusetts, 1968.

Williams, R. A., "A Study of the Drilling Process," *J. Eng. Ind.*, 1973.

5

Tool Wear and Affiliated Production Costs

5.1 INTRODUCTION

Optimization of machining conditions is a primary concern in the economic analysis of metal cutting operations. For the manufacturer, the goal is to produce parts that are within specifications in such a fashion as to make the endeavor profitable. This implies that there is an effort to produce satisfactory parts at the lowest possible cost.

In order to do this, the performance of tooling and machining conditions needs to be scrutinized. Proper analysis can set in balance the relationship between minimizing production costs and maximizing production rates. High-profitability operating conditions can often be attained through manufacturing analyses that lead to changes in production techniques that utilize new developments in the state of the art.

As an example, it can readily be shown that large increases in machining rates have been achieved by a change in a specific tool material or by changing an operation from dry machining to one using a cutting fluid. The application of nontraditional machining processes have also had a dramatic impact on manufacturing rates. New cutting materials as well as manufacturing processes are continuously evolving. The truism that the product can be produced less expensively by an improved technique can be substantiated by a comparison of production methods and costs spanning a long period of time. Production rates have been continuously growing while relative costs have been dramatically declining.

If a tool were available that did not wear out, the task of increasing production and decreasing costs would be simplified. By increasing the speeds of operation, more parts would be produced and the time affiliated with producing each part would be reduced. However, in the real world, metal cutting tools do wear out. They wear out at a more rapid rate when the cutting speed is increased. Higher speeds can lead to tool wear rates that can reduce production rates and increase production costs. For this reason, the availability of analytical expressions that accurately describe the relationship between tool wear and cutting conditions is important.

Through the analysis of the production rate, cutting conditions can be idealized to yield what may be considered the maximum production rate. This is a combination of machine settings that produce the most parts per unit of time. Of interest is that any change (increase or decrease) of cutting conditions from the so-called idealized case will reduce the production rate.

In a similar fashion, with the availability of an analytical expression relating tool wear with cutting conditions, an idealization can be made to derive an expression that will dictate the minimum-cost case. This is a machine setting that will produce parts at the lowest cost. Any change (increase or decrease) from this condition will have a tendency to increase the cost of the operation.

A comparison of results from the maximum-production case with those from the minimum-cost case can prove to be of interest. These two cases provide boundaries for a range within which one can balance high production with low costs of operation.

Another topic in this chapter deals with comparative costs of different methods of tool use. Emphasis is placed on how a tool is used. It is not as important to have good tooling available as it is to use the good tooling to the advantage of high production rates and low costs. In other words, it is not what you have, but rather what you do with what you have, that counts.

The final topic in this chapter deals with the use of money. Financial considerations are analyzed in light of the time rate affiliated with loans. Break-even points and financial gains are illustrated with numerical examples.

5.2 DERIVATION OF TAYLOR'S TOOL LIFE EQUATION

In order for a tool to successfully perform under the harsh environment of high temperatures and high forces involved with the cutting process, it must sacrifice itself in terms of wear. As the tool wears, eventually it fails to perform its intended objective of cutting. A broad description of tool failure can be given within the context that the tool loses its capability to produce parts within given specifications. These specifications could include surface quality, dimensional stability, and production rates.

More violent tool failures involve a complete erosion of the cutting edge. Under this circumstance, cutting ceases and rubbing action commences. High forces usually develop. When this happens, the work performed is converted to heat, and high temperatures ensue. This can lead to what is referred to as a *catastrophic tool failure*, when the cutting edge collapses completely. The result is a rubbing action between the tool and work material, generating a large increase in the the forces acting on the tool.

Tool wear is an intricate phenomenon influenced by a multitude of factors; among these are: abrasion, adhesion, diffusion, and plastic deformation. With abrasion, hard portions of the workpiece slide over the tool, causing wear by scraping a part of the tool away. With adhesion, a part of the tool forms a bond with the chip and is carried away, causing wear to occur. With diffusion, there is a dissolving of the tool due to conditions of high stress and high temperature that causes a weakening in the bond of the tool particles. Plastic deformation as a mechanism of tool wear is caused by high-stress and high-temperature conditions. The tool is placed into a condition where it plastically deforms, causing the cutting edge to weaken.

Figure 5.1 illustrates the nose, flank, and face wear regions on a turning tool. As the tool plows through the workpiece during the cutting process, it wears along its nose and flank in a fashion that has a tendency to erode the clearance angles. As shown in Fig. 5.1, the face wear is in the form of a cratering action.

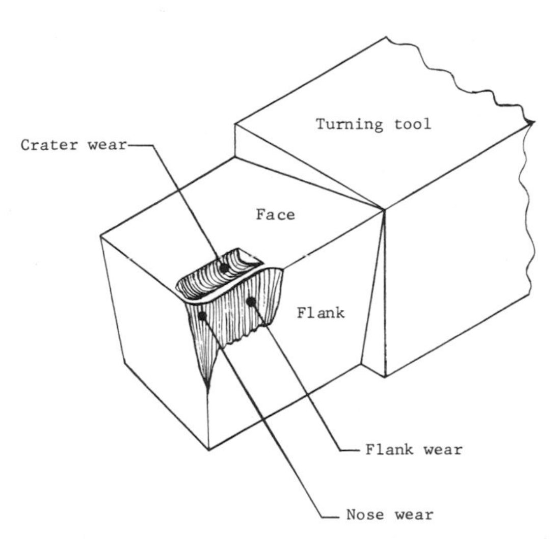

Fig. 5.1. Nose, flank, and face wear regions on a turning tool.

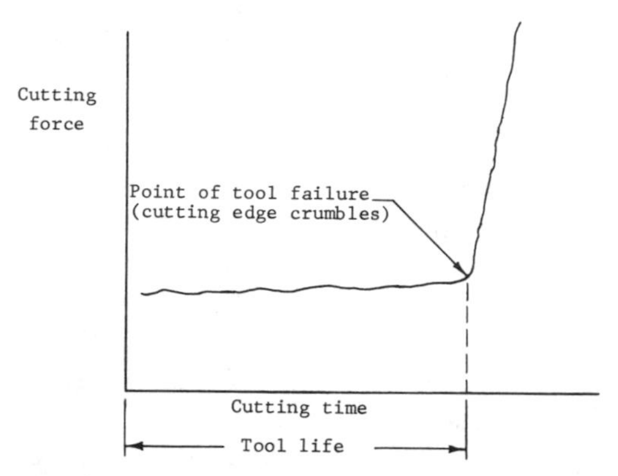

Fig. 5.2. Representation of relationship between cutting force and cutting time leading to catastrophic tool failure.

In the case where the cutting edge softens due to the high temperature and high stress generated by the cutting process, the result can lead to a plastic deformation of the tool, causing a crumbling of the cutting edge. A fracture or chipping on the cutting edge due to heavy loads or hard particles in the workpiece can also produce a crumbling of the cutting edge. If one were to monitor the magnitude of the cutting force as a function of cutting time, a drastic increase in the cutting force would occur at the time of the collapse of the cutting edge. It would then be obvious that the tool had failed. Figure 5.2 illustrates a case of this type, where the tool life is set at the collapse of the cutting edge. As can be seen, the cutting force rapidly increases in magnitude beyond the point labeled tool life.

A tool may also be considered to have failed if it is worn to a predetermined wear limit. This is a convenient way of measuring tool life. Other methods of identifying tool life may in terms of time of actual operation, number of pieces machined, or even in volume of metal removed.

In 1906, Frederick W. Taylor[1] proposed a relationship between cutting speed and tool life. It is of the form

$$V \times T^n = C \tag{5.1}$$

where V is the cutting speed, T is the tool life, and n and C are constants. Equation 5.1 received historic acceptance and is referred to as

[1] F. W. Taylor, *On the Art of Cutting Metals*, American Society of Mechanical Engineers, New York, 1906, p. 159.

Taylor's tool life equation. The constants of the equation can be determined for a particular cutting operation by conducting a series of experiments from which measurements of tool wear can be taken. However, before any test can be conducted, some criterion must be chosen to define tool life.

As an example, a flank wear limit of some value can cause undesirable cutting conditions. If that is the case, then this tool wear value can be set as the limit of tool wear at which point the tool is considered to fail. Measurements of flank wear can easily be made with the aid of a microscope. Figure 5.3 illustrates the flank wear on a turning tool.

With the feed and depth of cut held constant, the experimental data enables one to plot a series of curves relating tool wear with cutting time. Figure 5.4 shows three example curves for three different settings of the cutting speed. Experimental points are shown with a smooth curve drawn through the points. Cutting speed 1 (V_1) is larger than cutting speed 2 (V_2), which is larger than cutting speed 3 (V_3). As can be seen, it takes more time to reach the wear limit (W_r) with the lower cutting speed (V_3) than with the higher cutting speed (V_1).

A predetermined flank wear (W_r) is set as the wear limit. When the wear reaches this value, the tool is assumed to fail, that is, it is found not to function to desired expectations. The intersection of the wear limit line with the wear curve determines the tool life for a cutting speed setting. From these data, Eq. 5.1 can now be expressed as

$$V_1 \times (T_1)^n = V_2 \times (T_2)^n = V_3 \times (T_3)^n = C \tag{5.2}$$

The experimental relationship between the cutting speed and the tool life represented in Fig. 5.4 can be plotted on rectangular coordinate

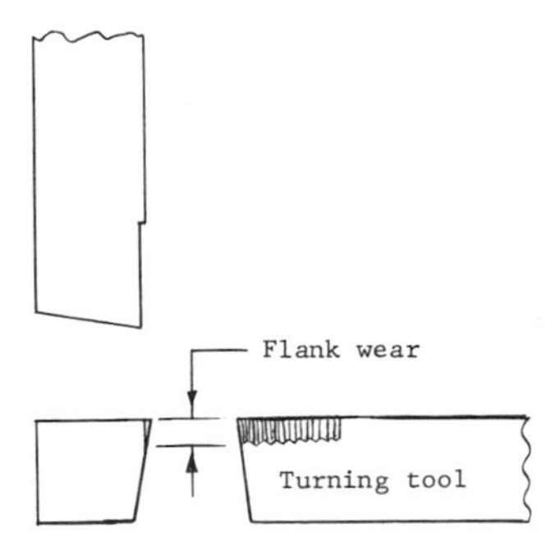

Fig. 5.3. Three views of turning tool with illustration of flank wear.

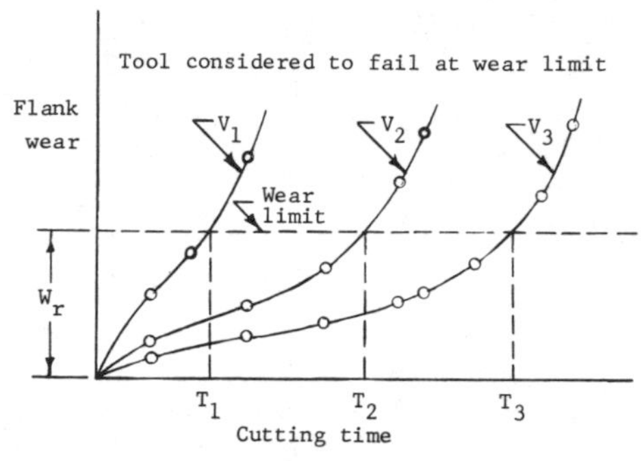

Fig. 5.4. Relationship between flank wear and cutting time for three settings of cutting speed.

graph paper. This is shown in Fig. 5.5. As can be seen, the curve is of an exponential shape.

If the experimental relationship between cutting speed and tool life, as determined from Fig. 5.4, is plotted on log–log graph paper, the points fall on an approximate straight line, as shown in Fig. 5.6. The equation of a straight line in rectangular coordinates is of the form

$$y = b + ax$$

where b is the y-intercept and a is the slope of the curve. On log–log paper, the experimental data expressed as an equation of a straight line would be of the form

$$\log y = \log b + a \log x$$

or

$$\log V = \log C - n \log T \quad (5.3)$$

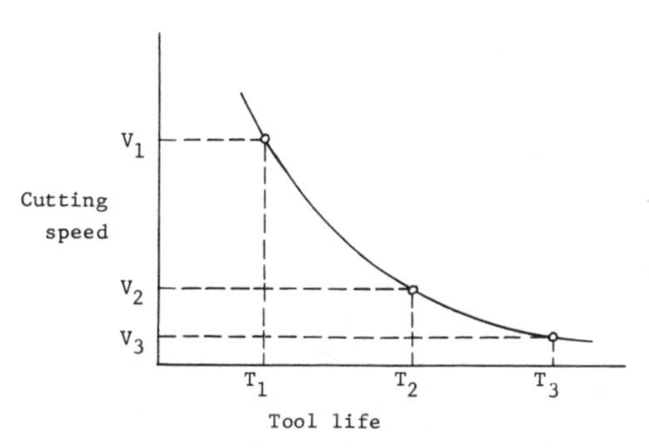

Fig. 5.5. Plot of experimental data on rectangular coordinate graph paper.

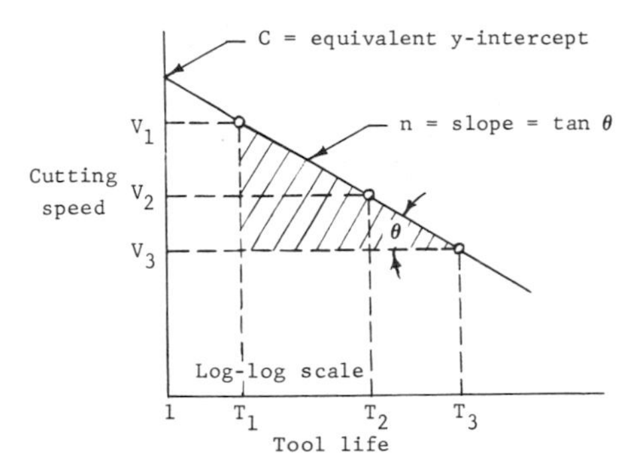

Fig. 5.6. Plot of experimental data on log–log graph paper.

This satisfies the expression

$$V \times T^n = C \tag{5.1}$$

where n is the slope of the curve on log–log paper and C is the equivalent y-intercept on log–log paper. From Fig. 5.6, the slope of the straight line can be expressed as

$$n = \tan\theta = \frac{\log V_1 - \log V_3}{\log T_3 - \log T_1} \tag{5.4}$$

The equivalent y-intercept, which is the value of the constant C, can be extracted from Eq. 5.3. When $\log T = 0$, a case where $T = 1$, then from Eq. 5.3 we obtain

$$\log V = \log C - n(0)$$

and

$$C = V$$

The value of C can also be extracted directly from Fig. 5.6 by noting where the straight line intersects the tool life value of 1 min. A confirmation of the value of the constant C can be made by using Eq. 5.1. By substituting the value of 1 for T, we obtain

$$C = V \qquad \text{for } T = 1$$

since 1 raised to any power is equal to 1. Analytically, the value of the constant C can also be evaluated from experimental data by noting that

$$C = V_1 \times (T_1)^n = V_2 \times (T_2)^n = V_3 \times (T_3)^n \tag{5.2}$$

The procedure of extracting the constants of Taylor's tool life equation for a cutting operation is demonstrated by using the experimental data listed in Table 5.1. An alloy steel is being machined with a carbide

Table 5.1. Tool Wear Experimental Data

	Cutting Speed		Time	Flank Wear	
Test	(ft/min)	(m/min)	(min)	(in.)	(mm)
1	600	183	2	0.014	0.356
1	600	183	4	0.025	0.635
2	500	152	2	0.007	0.178
2	500	152	4	0.012	0.305
2	500	152	6	0.018	0.457
2	500	152	8	0.026	0.660
3	400	122	2	0.004	0.102
3	400	122	4	0.007	0.178
3	400	122	6	0.010	0.254
3	400	122	8	0.012	0.305
3	400	122	12	0.015	0.356
3	400	122	15	0.018	0.457
3	400	122	20	0.029	0.737
4	300	91.4	6	0.005	0.127
4	300	91.4	12	0.008	0.203
4	300	91.4	25	0.011	0.279
4	300	91.4	35	0.013	0.330
4	300	91.4	40	0.015	0.381
4	300	91.4	45	0.017	0.432
4	300	91.4	50	0.020	0.508
4	300	91.4	55	0.023	0.584

tool. Machine settings are: depth of cut, 0.100 in. (2.54 mm); feed, 0.007 in./rev (0.178 mm). It is found that when the flank wear reaches a value of 0.020 in. (0.508 mm), the tool no longer performs satisfactorily and is assumed to have failed. A graphical representation of the experimental data is shown in Fig. 5.7.

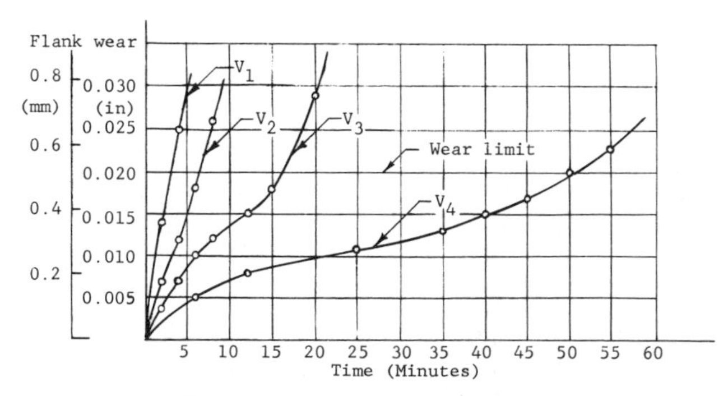

Fig. 5.7. Graphical representation of experimental data.

Table 5.2. Cutting-Speed-Tool Life Data

Cutting Speed		Tool Life
(ft/min)	(m/min)	min
600	182.9	3.16
500	152.4	6.55
400	121.9	16.00
300	91.44	50.50

By noting, in Fig. 5.7, the intersection of the wear curves with the wear limit line, the relationship between cutting speed and tool life can be established. Table 5.2 lists this relationship. A plot, in rectangular coordinates of the data listed in Table 5.2, is shown in Fig. 5.8. When these data are plotted on log–log paper, as shown in Fig. 5.9, the experimental points have a tendency to line up and a straight line can be drawn. The constants for Taylor's tool life equation can be determined by examining Fig. 5.9. The value of the constant C, which is the equivalent y-intercept (cutting speed at tool life of 1 min), is found to be 800. The value of the constant n is equal to the slope of the straight line on the log–log plot. The angle θ is measured to be 14°. The tangent of this angle is equal to 0.25, which is the value of n for this cutting operation. As a result, Taylor's tool life equation can now be written as

$$V \times T^{0.25} = 800 \tag{5.5}$$

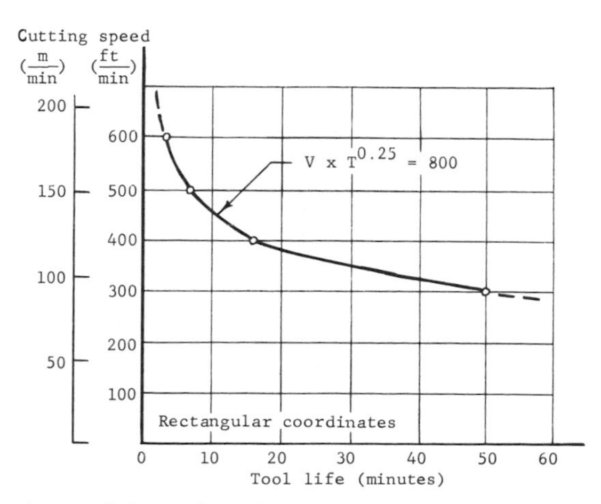

Fig. 5.8. Experimental data plotted on rectangular coordinate graph paper.

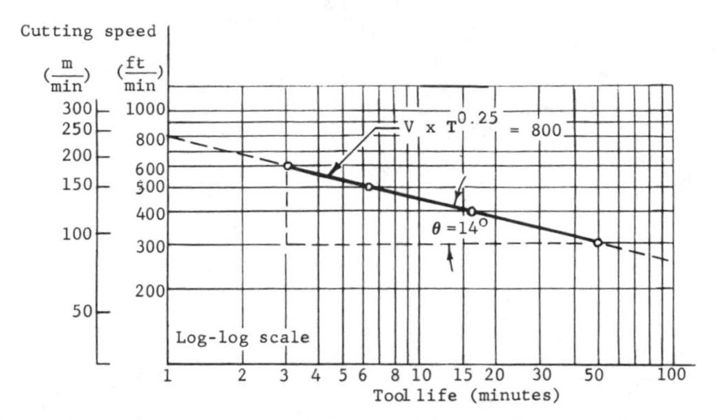

Fig. 5.9. Experimental data plotted on log–log graph paper.

The value of the constant n can be confirmed through the use of Eq. 5.4. Taking values from Table 5.2 yields

$$n = \frac{\log V_1 - \log V_4}{\log T_4 - \log T_1} = \frac{\log 600 - \log 300}{\log 50.5 - \log 3.16}$$

$$= \frac{2.78 - 2.48}{1.703 - 0.500} = 0.25$$

The value of the constant C can be confirmed by taking values from Table 5.2 and substituting into Eq. 5.2. To illustrate:

$$C = 600 \times (3.16)^{0.25} = 500 \times (6.55)^{0.25}$$
$$= 400 \times (16)^{0.25} = 300 \times (50.5)^{0.25}$$
$$= 800$$

A point of interest is that caution should be exercised when using tool life equations beyond the operating region in which they were derived. In Fig. 5.9, dashed lines indicate the region beyond the experimental points. In actual practice, when tool life points are placed on a log–log plot, they fall within a range of distribution. They may be linear over a relatively short range of cutting speed, but have a tendency to vary from the linear relationship over a wide range of cutting speeds.

If the linear relationship is assumed to exist, then two tool life tests of a metal cutting operation can establish the value of the constants in Taylor's tool life equation. Example 5.1 illustrates an application of this type.

EXAMPLE 5.1. It is desired to set a cutting speed on a machine tool to have a tool life of 30 min. The work material is AISI-1045 steel and the tool material is carbide. The feed is set at 0.010 in. (0.254 mm) and the depth of cut is 0.200 in. (5.08 mm). When the machine is set to run at 650 ft/min (198 m/min), the tool life is found to be 12 min. When the machine is set to run at 450 ft/min (137 m/min), the tool life is found to be 55 min. Assuming that the cutting-speed–tool-life relationship is of the type $V \times T^n = C$, determine the recommended cutting speed for a tool life of 30 min.

Substituting numerical values into Eq. 5.1 yields

$$650 \times (12)^n = 450 \times (55)^n = C \tag{1}$$

Solving for the constant n, we obtain

$$\tfrac{650}{450} = (\tfrac{55}{12})^n$$

$$1.444 = (4.58)^n$$

$$\log 1.444 = n \times \log 4.58$$

or

$$0.1597 = n \times (0.6609)$$

$$n = 0.2416$$

The constant C can be determined by substituting into Eq. 1:

$$\begin{aligned} C &= 650 \times (12)^{0.2416} \\ &= 1184 \end{aligned}$$

As a result, Taylor's equation can be written as

$$V \times T^{0.2416} = 1184 \tag{2}$$

Solving for the recommended cutting speed for a tool life of 30 min by substituting into Eq. 2 gives

$$\begin{aligned} V &= \frac{1184}{(30)^{0.2416}} \\ &= 520 \text{ ft/min } (158 \text{ m/min}) \end{aligned}$$

Taylor's tool life equation can be expanded to include the effects of the feed (f) and the depth of cut (d) by writing that

$$V \times T^n = C = \frac{K}{f^{n_1} \times d^{n_2}}$$

or

$$V \times T^n \times f^{n_1} \times d^{n_2} = K \tag{5.6}$$

where K is a constant, n_1 is the exponent of feed, n_2 is the exponent of depth of cut, f is the feed, d is the depth of cut, and V is the cutting speed. Equation 5.6 represents a convenient assumed relationship, which in some cases can be considered to be valid over restricted ranges of the variables. Unfortunately, there is no strict adherence to the fairly simple relationship represented by Eq. 5.6. Therefore, care should be exercised in the application, and shop trials should be used for verification of calculations. Nevertheless, if the assumption is made that Eq. 5.6 is valid over the restricted range where it will be used, then numerical calculations can be performed.

In order to demonstrate the effect on tool life of changing different machine settings, let it be assumed that for an operation using a carbide tool, Eq. 5.5 can be expanded to be written as

$$V \times T^{0.25} = 800 = \frac{83.39}{f^{0.40} \times d^{0.12}}$$

or

$$V \times T^{0.25} \times f^{0.40} \times d^{0.12} = 83.39 \tag{5.7}$$

which is of the form

$$V \times T^{n} \times f^{n_1} \times d^{n_2} = K \tag{5.6}$$

The tool life can be isolated by writing

$$T^{n} = \frac{K}{V \times f^{n_1} \times d^{n_2}}$$

or

$$T = \frac{K^{1/n}}{V^{1/n} \times f^{n_1/n} \times d^{n_2/n}} \tag{5.8}$$

By substituting values from Eq. 5.7, Eq. 5.8 can be expressed as

$$T = \frac{48.36 \times 10^{6}}{V^{4} \times f^{1.6} \times d^{0.48}} \tag{5.9}$$

It is noted that the exponent of the cutting speed $(1/n)$ is larger than the exponent of the feed (n_1/n), which is larger than the exponent of the depth of cut (n_2/n). The significance of these differences is that the tool life is most affected by changes in the cutting speed, less affected by changes in the feed, and least affected by changes in the depth of cut. In other words, for a given metal volumetric removal rate in terms of tool wear, it is better to use a large depth of cut and a low cutting speed.

To demonstrate the influence on tool life of changing settings of cutting speed, feed, and depth of cut, the following example is given.

EXAMPLE 5.2. A metal cutting operation using a carbide tool is represented by Eq. 5.7. The cutting speed is 300 ft/min (91.44 m/min), the feed is 0.007 in. (0.178 mm), and the depth of cut is 0.100 in. (2.54 mm). It is desired to find the influence on the tool life that a 100% increase in the volumetric metal removal rate has when:

(a) Increasing the cutting speed by 100%.
(b) Increasing the feed by 100%.
(c) Increasing the depth of cut by 100%.

By substituting into Eq. 5.9, the tool life with the original machine settings can be written as

$$T = \frac{48.36 \times 10^6}{(300)^4 \times (0.007)^{1.6} \times (0.100)^{0.48}}$$
$$= 50.6 \text{ min}$$

If the cutting speed is increased to 600 ft/min (182.9 m/min), then

$$T = \frac{48.36 \times 10^6}{(600)^4 \times (0.007)^{1.6} \times (0.100)^{0.48}}$$
$$= 3.16 \text{ min}$$

If the feed is increased to 0.014 in./rev (0.356 mm/rev), then

$$T = \frac{48.36 \times 10^6}{(300)^4 \times (0.014)^{1.6} \times (0.100)^{0.48}}$$
$$= 16.68 \text{ min}$$

If the depth of cut is increased to 0.200 in. (5.08 mm), then

$$T = \frac{48.36 \times 10^6}{(300)^4 \times (0.007)^{1.6} \times (0.200)^{0.48}}$$
$$= 36.25 \text{ min}$$

As can be seen from the results in example 5.2, increases in the cutting speed as an alternative to increasing volumetric metal removal rates have a detrimental effect on tool life. When compared with the results of increases in the depth of cut, the difference is substantial.

5.3 IDEALIZATION OF CUTTING CONDITIONS FOR MAXIMUM PRODUCTION

With the availability of an analytical expression relating tool life and cutting speed, an analysis can be performed to examine production rates. A measure of the output of a single cutting tool edge can be made in terms of the quantity of material cut during the life of the tool. This can be expressed as

$$Q_t = 12 \times V \times f \times d \times T \tag{5.10}$$

where Q_t is the volume of material cut during life of a single cutting edge, V is the cutting speed, f is the feed, d is the depth of cut, and T is the tool life. Substituting for the cutting speed from Taylor's equation, where

$$V = \frac{C}{T^n}$$

then Eq. 5.10 can be written as

$$Q_t = 12 \times C \times f \times d \times T^{(1-n)} \tag{5.11}$$

The production rate (P_r) can be expressed in terms of the quantity of material cut and the affiliated time associated with the cutting process. This is of the form

$$P_r = \frac{Q_t}{\text{Time}}$$

where time is the tool life plus additional time affecting the production rate. When one considers the time directly associated with the production rate, the tool life time and the time required to change the tool are found to be important. These two times represent the time over which the quantity of material being cut is distributed. During the actual cutting, the tool life time is being absorbed. Once the tool wears out, it must be replaced. This tool change time is also charged against the production rate insofar as it affects the production rate. As a result, the production rate can be written as

$$P_r = \frac{Q_t}{T + t_1} \tag{5.12}$$

where t_1 is the tool change time. Substituting Eq. 5.11 into Eq. 5.12 yields

$$P_r = \frac{12 \times C \times f \times d \times T^{(1-n)}}{T + t_1} \tag{5.13}$$

From Taylor's equation, the tool life can be expressed as

$$T = \left(\frac{V}{C}\right)^{1/n} \tag{5.14}$$

The relationship between the production rate and the cutting speed, which is related to the tool life through Eq. 5.14, can be evaluated by substituting numerical values into Eqs. 5.13 and 5.14. Table 5.3 lists

Table 5.3. List of Production Rate Data

Cutting Speed		Tool Life	Production Rate	
(ft/min)	(m/min)	(min)	(in.^{3}/min)	(cm^3/sec)
200	61	256.00	1.67	0.455
300	91	50.50	2.42	0.662
400	122	16.00	2.99	0.815
500	152	6.55	3.22	0.879
600	183	3.16	3.09	0.842
700	213	1.71	2.71	0.739
800	244	1.00	2.24	0.612
900	274	0.62	1.79	0.489

production rates corresponding to cutting speed and tool life for the case where Taylor's equation is

$$V \times T^{0.25} = 800 \tag{5.15}$$

and

$$f = 0.007 \text{ in./rev } (0.178 \text{ mm/rev})$$

$$d = 0.100 \text{ in. } (2.54 \text{ mm})$$

$$t_1 = 2 \text{ min}$$

A graphical representation of the data in Table 5.3 is shown in Fig. 5.10. Substituting the given numerical values into Eqs. 5.13 and 5.14

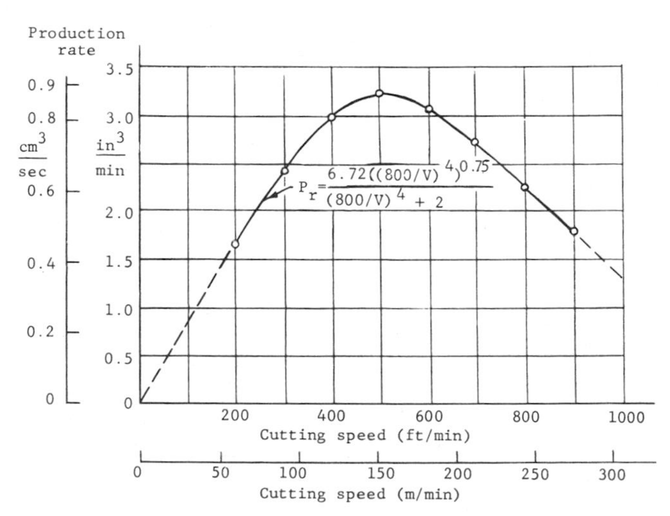

Fig. 5.10. Production rate versus cutting speed.

allows us to write the production rate and the tool life as

$$P_r = \frac{6.72 \times T^{0.75}}{T + 2} \tag{5.16}$$

$$T = \left(\frac{800}{V}\right)^4 \tag{5.17}$$

Combining Eqs. 5.16 and 5.17 yields

$$P_r = \frac{6.72[(800/V)^4]^{0.75}}{(800/V)^4 + 2} \tag{5.18}$$

Figure 5.10 graphically represents the relationship between the production rate and the cutting speed for the case described by Eq. 5.18. An examination of Fig. 5.10 reveals that the maximum production rate occurs when the cutting speed is in the vicinity of 500 ft/min (152.4 m/min).

To determine the value of the tool life for maximum production, Eq. 5.13 can be differentiated and the derivative (slope) can be set to zero. To illustrate, Eq. 5.13 can be written as

$$P_r = 12 \times C \times f \times d \times T^{(1-n)} \times (T + t_1)^{-1}$$

or

$$P_r = K_1 \times T^{(1-n)} \times (T + t_1)^{-1}$$

where

$$K_1 = 12 \times C \times f \times d$$

Differentiating yields

$$\begin{aligned} \frac{dP_r}{dT} &= K_1[T^{(1-n)}(-1)(T + t_1)^{-2} + (T + t_1)^{-1}(1 - n)T^{(1-n)-1}] \\ &= K_1\left[\frac{-T^{(1-n)}}{(T + t_1)^2} + \frac{(1 - n)}{(T + t_1)T^n}\right] \\ &= K_1\left[\frac{-T + (1 - n)(T + t_1)}{(T + t_1)^2 T^n}\right] \end{aligned} \tag{5.19}$$

When the slope dP_r/dT is equal to zero, the production rate as a function of the tool life is equal to a maximum value. Equation 5.19 will be equal to zero when

$$-T + (1 - n)(T + t_1) = 0 \tag{5.20}$$

The tool life for maximum production can be extracted from Eq. 5.20 by solving for T. As a result,

$$-T + T - nT + t_1 - nt_1 = 0$$

from which

$$nT = t_1(1 - n)$$

and

$$T = \frac{t_1(1 - n)}{n}$$

or

$$T_{mp} = t_1\left(\frac{1}{n} - 1\right) \tag{5.21}$$

Equation 5.21 is revealing insofar as it indicates that the tool life for maximum production for a given work material and tool is a function of the tool change time.

With the availability of Eq. 5.21, the cutting speed setting for maximum production can be evaluated. Using the case illustrated in Fig. 5.10, where

$$V \times T^{0.25} = 800$$

and

$$t_1 = 2 \text{ min}$$

the tool life for maximum production is

$$T_{mp} = t_1\left(\frac{1}{n} - 1\right) = 2\left(\frac{1}{0.25} - 1\right)$$
$$= 6 \text{ min}$$

Substituting into Taylor's tool life equation for the case represented by Eq. 5.15, the cutting speed for maximum production can be written as

$$V_{mp} = \frac{C}{(T_{mp})^n} = \frac{800}{6^{0.25}}$$
$$= 511 \text{ ft/min } (155.8 \text{ m/min})$$

EXAMPLE 5.3. Calculate the cutting speed for maximum production for the case of a high-speed tool machining a cast-iron workpiece.

Taylor's equation for the operation is of the form

$$V \times T^{0.12} = 180$$

and the tool change time is 1.5 min.

Solving for the tool life for maximum production by substituting into Eq. 5.21 yields

$$T_{mp} = t_1 \left(\frac{1}{n} - 1\right) = 1.5 \left(\frac{1}{0.12} - 1\right)$$

$$= 11 \text{ min}$$

Substituting T_{mp} into Taylor's life equation gives

$$V_{mp} = \frac{180}{T^{0.12}} = \frac{180}{(11)^{0.12}}$$

$$= 135 \text{ ft/min } (41.14 \text{ m/min})$$

5.4 IDEALIZATION OF CUTTING CONDITIONS FOR MINIMUM COST

With the availability of an analytical expression relating tool life and cutting speed, an analysis can be performed to examine the costs affiliated with the metal cutting operation. If a single-point tool is considered, it is found that the total cost of machining is made up of four separate costs. These include: the cutting cost or expenditure charge when the machine is cutting; the tool change cost; the tool costs (depreciation and regrinding cost); and the nonproductive cost such as that affiliated with loading, unloading, tool approach, overtravel, etc. Analytically, the total cost can be written as

$$C_T = C_c + C_h + C_t + C_n \tag{5.22}$$

where C_T is the total cost, C_c is the cutting cost, C_h is the tool change cost, C_t is the tool cost, and C_n is the nonproduction cost. The relationship between costs and cutting speed, as listed in Eq. 5.22, is illustrated in Fig. 5.11. The combined costs are shown in Fig. 5.12, demonstrating the relationship between the total cost and the cutting speed. As can be seen, the point where the slope of the total cost is equal to zero is the point of operation where the total cost is a minimum.

Analytical expressions for the different costs that make up the total cost can be developed. If the single-point tool is considered for the

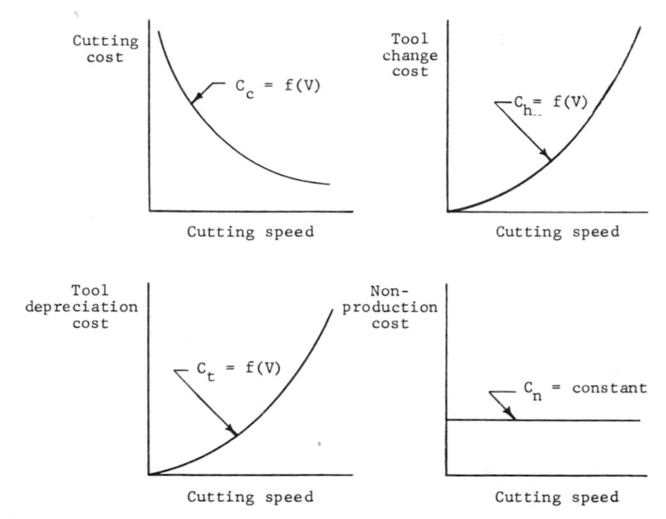

Fig. 5.11. Relationship between production costs and cutting speed.

turning operation, then the cutting cost per piece can be written as

$$\text{Cutting cost} = \text{Charge rate} \times \text{Cutting time}$$

where

$$\text{Cutting time} = \frac{\text{Length of cut}}{\text{Feed velocity}} = \frac{\text{Length of cut}}{\text{Feed} \times \text{rpm}}$$

$$= \frac{L}{f \times 12V/\pi D}$$

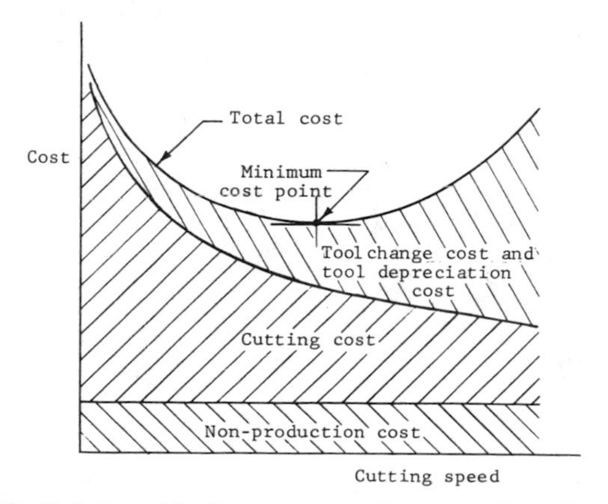

Fig. 5.12. Relationship between total cost and cutting speed.

As a result, the cutting cost can be written as

$$C_c = \frac{R_c \times L \times \pi \times D}{12 \times f \times V} \tag{5.23}$$

where R_c is the charge rate (includes overhead and related charges), L is the length of cut, D is the diameter of workpiece, f is the feed, and V is the cutting speed. The tool change cost can be written as

$$\text{Tool change cost} = \text{Charge rate} \times \begin{matrix}\text{Number of}\\ \text{tool failures}\\ \text{per piece}\end{matrix} \times \text{Tool change time}$$

where

$$\begin{matrix}\text{Number of}\\ \text{tool failures}\\ \text{per piece}\end{matrix} = \begin{matrix}\text{Time to machine}\\ \text{one part}\end{matrix} \times \begin{matrix}\text{Number of}\\ \text{tool changes}\\ \text{per unit time}\end{matrix}$$

$$\text{Time to machine one part} = \frac{\text{Length of cut}}{\text{Feed velocity}} = \frac{\text{Length of cut}}{\text{Feed} \times \text{rpm}}$$

$$= \frac{L}{f \times 12V/\pi D}$$

$$\text{Number of tool changes per unit time} = \frac{1}{\text{Tool life}} = \frac{1}{T}$$

$$\text{Time to change tool} = t_1$$

As a result,

$$\text{Tool change cost} = R_c \times \frac{L \times \pi \times D}{f \times 12V} \times \frac{1}{T} \times t_1$$

or

$$C_h = \frac{R_c \times L \times \pi \times D \times t_1}{12 \times f \times V \times T} \tag{5.24}$$

The tool depreciation cost can be written as

$$\text{Tool depreciation cost} = \begin{matrix}\text{Tool cost}\\ \text{per edge}\end{matrix} \times \begin{matrix}\text{Number of tool edges}\\ \text{used per workpiece}\end{matrix}$$

where

Tool cost per edge = C_e (includes original cost plus regrinding cost if any)

and

$$\text{Number of tool edges used per workpiece} = \frac{\text{Machining time}}{\text{Tool life}}$$

$$= \frac{\text{Length of cut}}{\text{Feed velocity} \times \text{tool life}}$$

$$= \frac{L}{f \times 12V/\pi D \times T} = \frac{\pi \times D \times L}{12 \times f \times V \times T}$$

Therefore,

$$C_t = \frac{C_e \times \pi \times D \times L}{12 \times f \times V \times T} \tag{5.25}$$

Finally, the nonproductive cost can be written in terms of time consumed in loading, unloading, approaching, possible overtravel and other noncutting time. In other words,

$$C_n = R_c \times t_n \tag{5.26}$$

where t_n is the nonproduction time. Taking Eqs. 5.23–5.26 into account, the total cost can be written as

$$C_T = C_c + C_h + C_t + C_n$$

or

$$C_T = \frac{R_c \times L \times \pi \times D}{12 \times f \times V} + \frac{R_c \times L \times \pi \times D \times t_1}{12 \times f \times V \times T} + \frac{C_e \times \pi \times D \times L}{12 \times f \times V \times T} + R_c \times t_n \tag{5.27}$$

With the availability of Eq. 5.27, a numerical example can now be used to illustrate the distribution of the individual costs that make up the total cost of machining a workpiece. For the sake of convenience by collecting terms, Eq. 5.27 can be written as

$$C_T = \left(\frac{\pi \times D \times L}{12 \times f \times V}\right)\left(R_c + \frac{R_c \times t_1}{T} + \frac{C_e}{T}\right) + (R_c \times t_n) \tag{5.28}$$

Cutting operation values to be used for this example are the same as those used in the evaluation of production rates. These are:

$$V \times T^{0.25} = 800$$

$$f = 0.007 \text{ in./rev } (0.178 \text{ mm/rev})$$

$$d = 0.100 \text{ in. } (2.54 \text{ mm})$$

$$t_1 = 2 \text{ min}$$

In addition, the diameter of the workpiece (D), the length of the cut (L), the charge rate (R_c), the cost per cutting edge (C_e), and the non-productive time (t_n) are given as

$$D = 3 \text{ in. } (76.2 \text{ mm})$$

$$L = 18 \text{ in. } (457.2 \text{ mm})$$

$$R_c = \$35/\text{hr} = \$0.583/\text{min}$$

$$C_e = \$1.75/\text{edge}$$

$$t_n = 1.8 \text{ min}$$

Table 5.4 lists the relationship between the costs and the cutting speed. The relationship between cost data and cutting speed is graphically represented in Fig. 5.13. An examination of Fig. 5.13 reveals that the minimum cost occurs when the cutting speed is in the vicinity of 400 ft/min.

EXAMPLE 5.4. A turning operation is in the process of being evaluated for costs. The following data are collected:

Cutting speed = 300 ft/min (91 m/min)

Tool life = 50.5 min

Feed = 0.007 in/rev (0.178 mm/rev)

Depth of cut = 0.100 in. (2.54 mm)

Diameter = 3 in. (76.2 mm)

Length of cut = 18 in. (457.2 mm)

Charge rate = \$35/hr = \$0.583/min

Table 5.4. List of Cost Data

Cutting Speed		Tool Life	C_T	C_c	C_h	C_t	C_n
(ft/min)	(m/min)	(min)	(\$)	(\$)	(\$)	(\$)	(\$)
100	30	4096	12.83	11.77	0.006	0.009	1.049
200	61	256.0	7.05	5.88	0.046	0.069	1.049
300	91	50.5	5.36	3.92	0.155	0.233	1.049
400	122	16.0	4.91	2.94	0.368	0.552	1.049
500	152	6.55	5.20	2.35	0.719	1.079	1.049
600	183	3.16	6.11	1.96	1.241	1.863	1.049
700	213	1.71	7.65	1.68	1.966	2.951	1.049
800	244	1.00	9.88	1.47	2.942	4.416	1.049
900	274	0.62	12.91	1.31	4.218	6.330	1.049

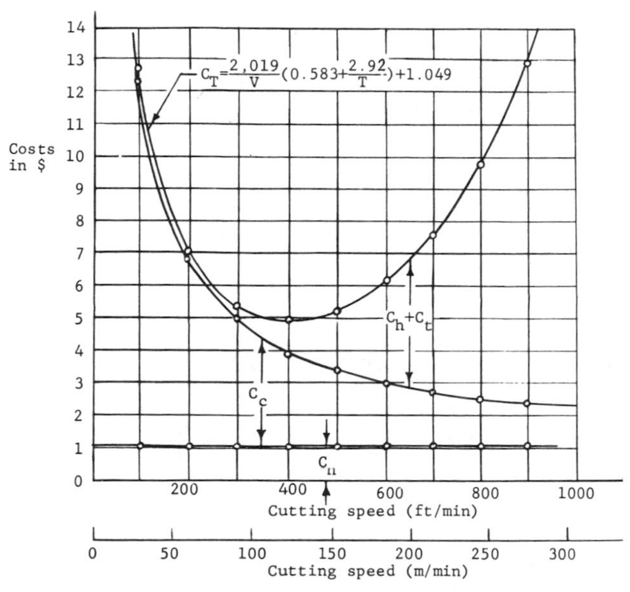

Fig. 5.13. Costs graphically displayed as a function of cutting speed.

Tool cost per edge = \$1.75

Nonproductive time/piece = 1.8 min

Tool change time = 2 min

From these data, calculate:

(a) The total cost.
(b) The cutting cost.
(c) The tool change cost.
(d) The tool depreciation cost.
(e) The nonproductive cost.

Solving for the total cost by substituting numerical values into Eq. 5.27 yields

$$C_T = \left(\frac{\pi \times D \times L}{12 \times f \times V}\right)\left(R_c + \frac{R_c \times t_1}{T} + \frac{C_e}{T}\right) + (R_c \times t_n)$$

$$= \left(\frac{3.14 \times 3 \times 18}{12 \times 0.007 \times 300}\right)\left(0.583 + \frac{0.583 \times 2}{50.5} + \frac{1.75}{50.5}\right) + (0.583 \times 1.8)$$

$$= \$5.36$$

The cutting cost can be evaluated by substituting into Eq. 5.23:

$$C_c = \frac{R_c \times L \times \pi \times D}{12 \times f \times V} = \frac{0.583 \times 18 \times 3.14 \times 3}{12 \times 0.007 \times 300}$$

$$= \$3.92$$

The tool change cost can be evaluated by substituting into Eq. 5.24:

$$C_h = \frac{R_c \times L \times \pi \times D \times t_1}{12 \times f \times V \times T} = \frac{0.583 \times 18 \times 3.14 \times 3 \times 2}{12 \times 0.007 \times 300 \times 50.5}$$

$$= \$0.155$$

The tool depreciation cost can be evaluated by substituting into Eq. 5.25:

$$C_t = \frac{C_e \times \pi \times D \times L}{12 \times f \times V \times T} = \frac{1.75 \times 3.14 \times 3 \times 18}{12 \times 0.007 \times 300 \times 50.5}$$

$$= \$0.233$$

Finally, the nonproduction cost can be evaluated by substituting into Eq. 5.25:

$$C_n = R_c \times t_n = 0.583 \times 1.8$$
$$= \$1.049$$

Taking into account the different costs of the operation, the total cost can also be determined by substituting into Eq. 5.22. As a result,

$$C_T = C_c + C_h + C_t + C_n$$
$$= 3.92 + 0.155 + 0.233 + 1.049$$
$$= \$5.36$$

EXAMPLE 5.5. For the operation described in example 5.4, it is desired to determine the position on the cost curve that corresponds to the cutting speed setting of 300 ft/min (91 m/min). The question being addressed is whether or not an increase in the cutting speed will decrease the cost of the operation. To perform the test, the cutting speed is increased to 400 ft/min (122 m/min), and it is found that the tool life correspondingly decreases to 16.0 min. From this information, calculate the total cost of the operation and determine the position on the cost curve that corresponds to the location of a cutting speed setting of 300 ft/min (91 m/min).

Using a technique similar to that applied in example 5.4 by substituting the numerical values for the test into Eq. 5.23–5.25 yields

$$C_T = C_c + C_h + C_t + C_n$$
$$= 2.94 + 0.368 + 0.552 + 1.049$$
$$= \$4.91$$

As can be seen from the results, the total cost is reduced. From this, the conclusion can be reached that the 300 ft/min (91 m/min) cutting speed setting is lower than the cutting speed for minimum cost operation. A comparison of the results is of interest. Although the tool change cost and the tool depreciation costs are much larger for the 400 ft/min (122 m/min) cutting speed setting, the total cost is lower because of the relatively large reduction in the cost of cutting. These values are graphically illustrated in Fig. 5.13.

To determine analytically the tool life that will yield a minimum-cost operation, a technique similar to that used in the maximum production case can be applied. From Eq. 5.28,

$$C_T = \left(\frac{\pi \times D \times L}{12 \times f}\right)\left(\frac{1}{V}\right)\left(R_c + \frac{R_c \times t_1}{T} + \frac{C_e}{T}\right) + (R_c \times t_n)$$

If

$$K_2 = \frac{\pi \times D \times L}{12 \times f \times C} \quad \text{and} \quad V = \frac{C}{T^n}$$

then

$$C_T = (K_2 \times T^n)[R_c + (R_c \times t_1)T^{-1} + (C_e \times T^{-1})] + (R_c \times t_n)$$

or

$$C_T = K_2R_cT^n + K_2R_ct_1T^{(n-1)} + K_2C_eT^{(n-1)} + R_ct_n$$

Differentiating yields

$$\frac{dC_T}{dT} = nK_2R_cT^{(n-1)} + (n-1)K_2R_ct_1T^{(n-2)} + (n-1)K_2C_eT^{(n-2)}$$

The minimum cost exists at the point where the slope is equal to zero, or

$$dC_T/dT = 0$$

Therefore,

$$0 = K_2T^{(n-1)}(nR_c + (n-1)R_ct_1T^{-1} + (n-1)C_eT^{-1})$$

or

$$\frac{nR_cT + (n-1)R_ct_1 + (n-1)C_e}{T} = 0$$

Solving for T yields

$$T = \frac{-(n-1)R_ct_1 - (n-1)C_e}{nR_c}$$

or

$$T = \left(\frac{1}{n} - 1\right) t_1 + \left(\frac{1}{n} - 1\right) \frac{C_e}{R_c}$$

$$T_{\mathrm{mc}} = \left(\frac{1}{n} - 1\right)\left(t_1 + \frac{C_e}{R_c}\right) \tag{5.29}$$

where T_{mc} is the minimum-cost tool life. Equation 5.29 indicates that for a given metal cutting operation with a certain work material being machined by a specific tool that is described by the constant n from Taylor's tool life equation, the minimum tool life is a function of the tool change time (t_1) as well as the equivalent time affiliated with tool depreciation (C_e/R_c).

EXAMPLE 5.6. Calculate the cutting speed and the tool life for minimum cost for the case illustrated in Fig. 5.13. Taylor's tool life equation is of the form

$$V \times T^{0.25} = 800$$

The tool change time is 2.0 min, the cost per cutting edge is \$1.75, and the charge rate is \$0.583/min.

Substituting into Eq. 5.29 yields

$$T_{\mathrm{mc}} = \left(\frac{1}{n} - 1\right)\left(t_1 + \frac{C_e}{R_c}\right)$$

$$= \left(\frac{1}{0.25} - 1\right)\left(2.0 + \frac{1.75}{0.583}\right)$$

$$= 15.0 \text{ min}$$

From Taylor's tool life equation,

$$V_{\mathrm{mc}} = \frac{C}{T^n} = \frac{800}{(15.0)^{0.25}}$$

$$= 406.5 \text{ ft/min } (123.9 \text{ m/min})$$

EXAMPLE 5.7. A turning operation is to be analyzed for a cutting speed setting at minimum cost. A high-speed tool is machining a low-carbon steel with a Bhn of 175. The tool life equation is of the form

$$V \times T^{0.12} = 180$$

Other factors of the operation are:

Feed = 0.008 in./rev (0.2032 mm/rev)

Depth of cut = 0.090 in. (2.286 mm)

Diameter = 4 in. (101.6 mm)

Length of cut = 10 in. (254 mm)

Tool change time = 3 min

Tool depreciation cost/edge = \$2.10

Charge rate = \$40/hr = \$0.667/min

Nonproduction time = 2 min

From this information calculate:

(a) Tool life for minimum cost.
(b) Cutting speed for minimum cost.
(c) Cutting cost.
(d) Tool change cost.
(e) Tool depreciation cost.
(f) Nonproduction cost.
(g) Total cost.

Solving for minimum cost tool life by substituting into Eq. 5.29 yields

$$T_{\mathrm{mc}} = \left(\frac{1}{n} - 1\right)\left(t_1 + \frac{C_e}{R_c}\right) = \left(\frac{1}{0.12} - 1\right)\left(3 + \frac{2.10}{0.667}\right)$$
$$= 45.09 \text{ min}$$

Solving for the cutting speed from the tool life equation gives

$$V_{\mathrm{mc}} = \frac{C}{T^n} = \frac{180}{45.09^{0.12}} = 114 \text{ ft/min (34.75 m/min)}$$

Solving for the cutting cost by substituting into Eq. 5.23 gives

$$C_c = \frac{R_c \times L \times \pi \times D}{12 \times f \times V_{\mathrm{mc}}} = \frac{0.667 \times 10 \times 3.14 \times 4}{12 \times 0.008 \times 114}$$
$$= \$7.65$$

Solving for the tool change cost by substituting into Eq. 5.24 gives

$$C_h = \frac{R_c \times L \times \pi \times D \times t_1}{12 \times f \times V \times T_{\mathrm{mc}}} = \frac{0.667 \times 10 \times 3.14 \times 4 \times 3}{12 \times 0.008 \times 114 \times 45.09}$$
$$= \$0.509$$

Solving for the tool depreciation cost by substituting into Eq. 5.25 gives

$$C_t = \frac{C_e \times \pi \times D \times L}{12 \times f \times V_{\mathrm{mc}} \times T_{\mathrm{mc}}} = \frac{2.10 \times 3.14 \times 4 \times 10}{12 \times 0.008 \times 114 \times 45.09}$$
$$= \$0.535$$

Solving for the nonproduction cost by substituting into Eq. 5.26 gives

$$C_n = R_c \times t_n = 0.667 \times 2$$
$$= \$1.334$$

Finally, solving for the total cost by substituting into Eq. 5.22 yields

$$C_T = C_c + C_h + C_t + C_n$$
$$= 7.65 + 0.509 + 0.535 + 1.334$$
$$= \$10.03$$

The total cost value can be confirmed by substituting numerical values into Eq. 5.28.

EXAMPLE 5.8. Using the data from example 5.7, a test is conducted to measure the influence on costs of a change in the cutting speed from the minimum-cost value. To measure this influence, the cutting speed is set above and below the minimum-cost value. In compliance with the tool life equation

$$V \times T^{0.12} = 180$$

a cutting speed of 125 ft/min (38.1 m/min) is found to yield a tool life of 20.85 min, whereas a cutting speed of 100 ft/min (30.48 m/min) is found to yield a tool life of 133.79 min. From this information determine the costs of operation for the two cutting speed settings.

For $V = 125$ and $T = 20.85$,

$$C_T = C_c + C_h + C_t + C_n$$
$$10.37 = 6.98 + 1.004 + 1.054 + 1.334$$

For $V = 100$ and $T = 133.79$,

$$C_T = C_c + C_h + C_t + C_n$$
$$10.46 = 8.727 + 0.1957 + 0.2054 + 1.334$$

As can be seen from example 5.8, the balance between cutting cost (C_c) and tool costs ($C_h + C_t$) is very sensitive with respect to the cutting speed.

5.5 EVALUATION OF MAXIMUM-PRODUCTION AND MINIMUM-COST OPERATIONS

With the availability of analytical expressions for maximum production and for minimum cost, the question arises as to which condition provides the most advantageous operation. If the cutting speed setting for maximum production and for minimum cost were the same value, then a setting at this level would produce the most parts for the lowest cost. However, this is not the case.

An examination of the tool life equations for maximum production and for minimum cost, namely,

$$T_{\mathrm{mp}} = t_1\left(\frac{1}{n} - 1\right) \tag{5.20}$$

$$T_{\mathrm{mc}} = \left(\frac{1}{n} - 1\right)\left(t_1 + \frac{C_e}{R_c}\right) \tag{5.29}$$

reveals that the difference between the equations is the term C_e/R_c. This term is equal to the equivalent time associated with tool depreciation. The influence of the term C_e/R_c is to give the tool life for minimum cost a higher value than that for maximum production. Consequently, when the cutting speed is evaluated through the tool life equation

$$V = \frac{C}{T^n}$$

the value of the cutting speed for minimum cost emerges as being lower than that for maximum production. This is illustrated in Fig. 5.14, which represents a combination of Figs. 5.10 and 5.13.

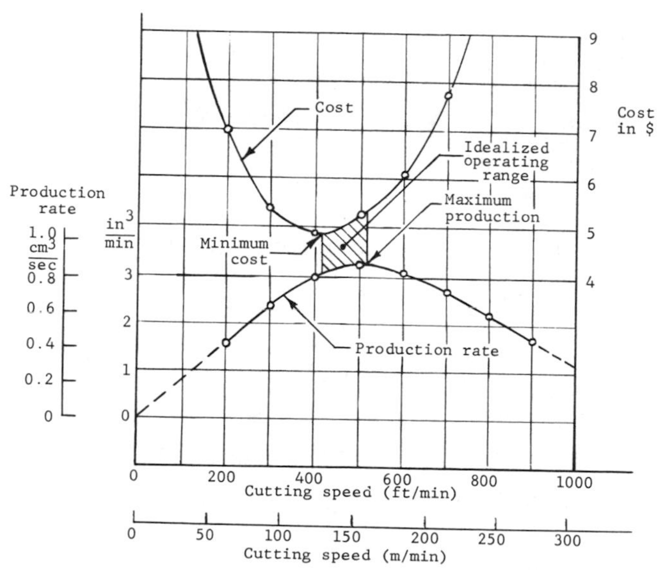

Fig. 5.14. Cost and production rate versus cutting speed.

From previous calculations with results shown in Fig. 5.14, the maximum production occurs at 511 ft/min (155.8 m/min), whereas the minimum cost occurs at 410.67 ft/min (125.17 m/min). The area between these cutting speeds can be called the *idealized operating range*. As can be seen in Fig. 5.14, within this range the cost is low and the production rate is high. Any shift within the range yields a gain in either higher production or in lower cost. However, this is not true beyond the idealized range. Any increase or decrease in cutting speed from the idealized range has the detrimental effect of simultaneously increasing cost and decreasing the production rate.

Between minimum cost and maximum production there is a partial beneficial effect from changes in the cutting speed. This is shown in Fig. 5.14, where an increase in cutting speed from the minimum-cost state increases production at the expense of increasing cost. Moving down from the maximum production state, a decrease in cutting speed decreases the cost to the minimum cost at the expense of a decrease in production. Beyond the idealized range as shown in Fig. 5.14, any change in cutting speed yields less production at a correspondingly higher cost.

A cutting speed reduction from the maximum production point provides for a decrease in cost to the point of minimum cost. This is accomplished at the expense of a decrease in production. This point is illustrated with a numerical example. The minimum cost can be calculated from Eq. 5.28 as

$$C_{\mathrm{Tmc}} = \left(\frac{\pi \times D \times L}{12 \times f \times V_{\mathrm{mc}}}\right)\left(R_c + \frac{R_c \times t_1}{T_{\mathrm{mc}}} + \frac{C_e}{T_{\mathrm{mc}}}\right) + (R_c \times t_n)$$

$$= \left(\frac{3.14 \times 3 \times 18}{12 \times 0.007 \times 406.5}\right)\left(0.583 + \frac{0.583 \times 2}{15} + \frac{1.75}{15}\right) + (0.583 \times 1.8)$$

$$= \$4.91$$

The corresponding production rate from Eq. 5.18 is

$$P_{\mathrm{rmc}} = \frac{12 \times C \times f \times d \times (T_{\mathrm{mc}})^{1-n}}{T_{\mathrm{mc}} + t_1}$$

$$= \frac{12 \times 800 \times 0.007 \times 0.100 \times (15)^{0.75}}{15 + 2}$$

$$= 3.01 \text{ in.}^3\text{/min } (0.822 \text{ cm}^3\text{/sec})$$

This compares with the maximum production rate of

$$P_{\mathrm{rmp}} = \frac{12 \times C \times f \times d \times (T_{\mathrm{mp}})^{1-n}}{T_{\mathrm{mp}} + t_1}$$

$$= 3.22 \text{ in.}^3\text{/min } (0.879 \text{ cm}^3\text{/sec})$$

As can be seen, operating at the minimum-cost cutting speed yields 0.21 in.^{3}/min (0.064 cm^3/sec) less than the maximum production rate. This represents an approximate 6% reduction from the maximum production rate.

In a similar fashion, the cost at maximum production can be analyzed. At the maximum-production-rate cutting speed of 511 ft/min (155.8 m/min), the corresponding tool life is 6 min. The cost of production at maximum production can be evaluated by substituting into Eq. 5.28:

$$C_{\mathrm{Tmp}} = \left(\frac{\pi \times D \times L}{12 \times f \times V_{\mathrm{mp}}}\right)\left(R_c + \frac{R_c \times t_1}{T_{\mathrm{mp}}} + \frac{C_e}{T_{\mathrm{mp}}}\right) + (R_c \times t_n)$$

$$= \left(\frac{3.14 \times 3 \times 18}{12 \times 0.007 \times 511}\right)\left(0.583 + \frac{0.583 \times 2}{6} + \frac{1.75}{6}\right) + (0.583 \times 1.8)$$

$$= \$5.27$$

The cost at maximum production is $0.36 higher than that for minimum cost. This is 7% higher than the cost for the minimum-cost operation.

If the assumption is made that the analysis is reliable, conclusions can be drawn. The most significant one is that the idealized operating range between minimum cost and maximum production involves a region of low cost and high production. Any operation beyond the idealized range, either higher or lower, has the detrimental effect of not only reducing production but also of increasing costs. In other words, beyond the idealized operating range it costs more to produce less.

The conversion of the production rate from units of volumetric rate to that of parts per hour can be determined by taking into account the volume of material removed per part during the period that includes the cutting time and the tool change time. In addition, the nonproductive time must also be taken into account. The nonproductive time is consumed in loading, unloading, approaching, possible overtravel, etc. When it is considered, its influence is to reduce the parts-per-hour production from that which takes into account only the volumetric rate of production.

Parts produced per hour with continuous cutting and tool changing, neglecting the nonproductive time, can be written as

$$P_c = \frac{60 \times P_r}{V_L} \tag{5.30}$$

where P_r is the volumetric production rate, V_L is the volume of material removed per part, and 60 is the conversion factor from minutes to hours. The influence of the nonproductive time on the production rate can be expressed in terms of

$$F_n = \frac{60}{60 + t_p} \tag{5.31}$$

where F_n is the nonproductive time factor and t_p is the nonproductive time used during 60 min of continuous cutting and tool changing. The time (t_p) can be written as

$$t_p = \left(\frac{60 \times P_r}{V_L}\right) t_n \tag{5.32}$$

where t_n is the nonproductive time per part. By combining Eq. 5.30–5.32, the production rate in parts per hour can now be written as

$$\begin{aligned} P_h &= \left(\frac{60 \times P_r}{V_L}\right) F_n \\ &= \left(\frac{60 \times P_r}{V_L}\right)\left(\frac{60}{60 + t_p}\right) \\ &= \left(\frac{60 \times P_r}{V_L}\right)\left[\frac{60}{60 + (60 \times P_r/V_L)t_n}\right] \\ &= \left(\frac{60 \times P_r}{V_L}\right)\left[\frac{V_L}{V_L + (P_r \times t_n)}\right] \\ &= \frac{60 \times P_r}{V_L + (P_r \times t_n)} \end{aligned} \tag{5.33}$$

For a turning operation, the volume per cut can be expressed in terms of the diameter, depth of cut, and length of cut, shown in Fig. 5.15 as

$$V_L = \frac{\pi}{4}\left[(D_b)^2 - (D_a)^2\right] \times L \tag{5.34}$$

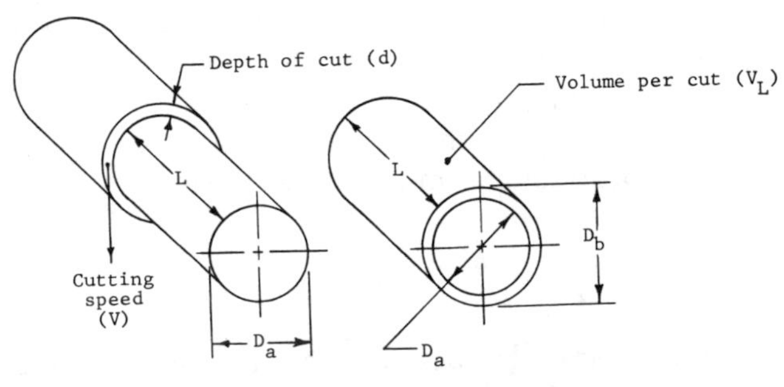

Fig. 5.15. Turning operation.

where D_b is the part diameter before cut, D_a is the part diameter after cut, L is the length of cut, $D_a = D_b - 2d$, and d is the depth of cut. The application of Eqs. 5.32 and 5.33 is illustrated by the following turning example.

EXAMPLE 5.9. It is desired to find the value of the production rate in parts per hour for the maximum production case illustrated in Fig. 5.14. The maximum production rate in terms of volume per unit time, with nonproductive time neglected, is 3.22 in.^{3}/min (0.879 cm^3/sec); the diameter of the workpiece is 3 in. (76.2 mm); the length of the cut is 18 in. (457.2 mm); the depth of cut is 0.100 in. (2.54 mm); and the nonproductive time per part is given as 1.8 min.

From the given data, calculate:

(a) The volume removed by the cut (V_L).
(b) The number of parts produced per hour with continuous cutting and tool changing, neglecting the nonproductive time (P_c).
(c) The nonproductive time used during 60 min of continuous cutting and tool changing (t_p).
(d) The nonproductive time factor (F_n).
(e) The production rate in parts per hr (P_p).

The volume removed by the cut is determined by substituting numerical values into Eq. 5.34:

$$V_L = \frac{\pi}{4}[(D_b)^2 - (D_a)^2] \times L$$

$$= \frac{3.14}{4}(3.0^2 - 2.8^2) \times 18$$

$$= 16.4 \text{ in.}^3 \ (268.6 \text{ cm}^3)$$

The number of parts per hour with continuous cutting and tool changing is evaluated from Eq. 5.30:

$$P_c = \frac{60 \times P_r}{V_L} = \frac{60 \times 3.22}{16.39}$$

$$= 11.78 \text{ parts/hour}$$

The nonproductive time used during 60 min of continuous cutting and tool changing is determined from Eq. 5.32:

$$t_p = \left(\frac{60 \times P_r}{V_L}\right) t_n = 11.78 \times 1.8$$

$$= 21.2 \text{ min}$$

The nonproductive factor evaluated from Eq. 5.31 is

$$F_n = \frac{60}{60 + t_p} = \frac{60}{60 + 21.2}$$
$$= 0.739$$

Finally, the production rate in parts per hour is determined from Eq. 5.33:

$$P_h = \frac{60 \times P_r}{V_L + (P_r \times t_n)} = \frac{60 \times 3.22}{16.4 + (3.22 \times 1.8)}$$
$$= 8.71 \text{ parts/hr}$$

Another technique by which the production rate in parts per hour can be evaluated is through an analysis of the time involved in the production of one part. This time involves the sum total of the machining time per part (t_m), the tool change time per part (t_c), and the nonproductive time per part (t_m).

The machining time is the time that it used during the actual cutting of the metal. It can be written as

$$T_m = \frac{L}{f \times \text{rpm}} \tag{5.35}$$

where T_m is the machining time per part, L is the length of cut, f is the feed, and rpm represents revolutions per minute. The tool change time per part is the time used on one part to change the tool because of the wear of the tool. It can be written as

$$T_{\text{cp}} = T_m \times \frac{1}{T} \times t_c$$

or

$$T_{\text{cp}} = \frac{L \times t_c}{f \times \text{rpm} \times T} \tag{5.36}$$

where T_m is the machining time per part, T is the tool life, and t_c is the tool change time per cutting edge. The nonproductive time is a constant for a particular workpiece. It can be expressed as

$$t_n = \text{constant} \tag{5.37}$$

By combining Eqs. 5.35–5.37, the time used by one part can be written as

$$T_p = T_m + T_{\text{cp}} + t_n \tag{5.38}$$

With the availability of Eq. 5.38, the production rate in terms of parts per hour can now be expressed as

$$P_h = \frac{60}{T_p} = \frac{60}{T_m + T_{cp} + t_n} \tag{5.39}$$

EXAMPLE 5.10. It is desired to confirm the production rate in parts per hr for the case described in example 5.9. The length of cut is 18 in. (45.72 cm), the diameter is 3 in. (76.2 mm), the feed is 0.007 in./rev (0.178 mm/rev), the cutting speed for maximum production is 511 ft/min (155.8 m/min), the corresponding tool life is 6 min, the tool change time is 2 min, and the nonproductive time is 1.8 min.

From the data, determine:

(a) The rpm of the workpiece.
(b) The machining time per part.
(c) The tool change time per part.
(d) The time used by one part.
(e) The production rate in parts per hr.

The rpm can be written as

$$\text{rpm} = \frac{12 \times V}{\pi \times D} = \frac{12 \times 511}{3.14 \times 2.9}$$
$$= 673 \text{ rev/min}$$

The machining time per part is determined by substituting into Eq. 5.36:

$$T_m = \frac{L}{f \times \text{rpm}} = \frac{18}{0.007 \times 673}$$
$$= 3.821 \text{ min}$$

The tool change time per part can be evaluated from Eq. 5.36 as

$$T_{cp} = \frac{T_m \times t_c}{T} = \frac{3.821 \times 2}{6}$$
$$= 1.274 \text{ min}$$

With the nonproductive time per part given as 1.8 min, the time used to complete one part can now be evaluated from Eq. 5.38 as

$$T_p = T_m + T_{cp} + t_n$$
$$= 3.821 + 1.274 + 1.8$$
$$= 6.895 \text{ min}$$

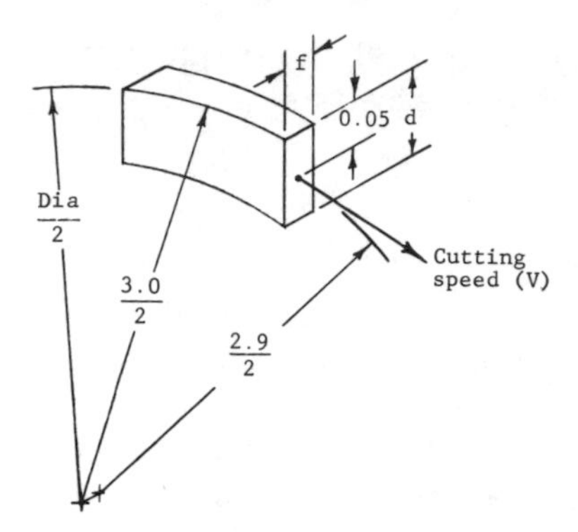

Fig. 5.16. Diagram of volumetric rate for turning example.

The production rate in terms of parts per hour can now be determined from Eq. 5.39 as

$$P_h = \frac{60}{T_p} = \frac{60}{6.895}$$

$$= 8.7 \text{ parts/hr.}$$

As can be seen, the results of example 5.9 are confirmed. Of special interest is the time distribution as indicated in example 5.10. When the different times are compared, it is found that the machining time exceeds the sum of the tool change and nonproduction time.

Another point of interest is that when the turning example is considered, the location of the cutting speed vector is of importance. It must be placed in such a fashion as to generate the same volume that a model has without any curvature. Figure 5.16 demonstrates this point and justifies using the diameter of 2.9 in. (73.66 mm) in example 5.10 rather than the outside diameter of 3.0 in. (76.2 mm).

5.6 COMPARISON OF TWO TOOLS FOR COST AND PRODUCTION

When a tool is evaluated for performance, the production cost per workpiece is usually the criterion used as a measure of performance. The cost, in turn, is directly affected by the life of the tool. When two tools are compared for performance, the life of the tool is the factor that plays a significant role in terms of the cost of the operation. The tool life also plays an important role as a controlling factor with regard to the production rate.

To show how important the tool life can be, a comparison is presented between two tools where the only difference in the operation is restricted to the life of the tool. The two tools to be considered have different tool life equations. Other factors such as tool change time, tool cost per edge, feed, depth of cut, work material, and nonproductive time per piece are all set to be the same for the two tools.

The first tool to be used in the comparison will be the one used in Section 5.4, which had a tool life equation of

$$V \times T^{0.25} = 800$$

The second tool will be one that has a tool life equation of

$$V \times T^{0.19} = 600 \tag{5.40}$$

Figure 5.17 illustrates the relationship between the two tools, taking into account the cutting speed and tool life. Using the same data as in Section 5.4 where

Feed (f) = 0.007 in./rev (0.178 mm/rev)

Depth of cut (d) = 0.100 in. (2.54 mm)

Tool change time (t_1) = 2 min

Part diameter (D) = 3 in. (76.2 mm)

Length of cut (L) = 18 in. (457.2 mm)

Charge rate (R_c) = \$35/hr = \$0.583/min

Cost per cutting edge (C_e) = \$1.75/edge

Nonproductive time (t_n) = 1.8 min

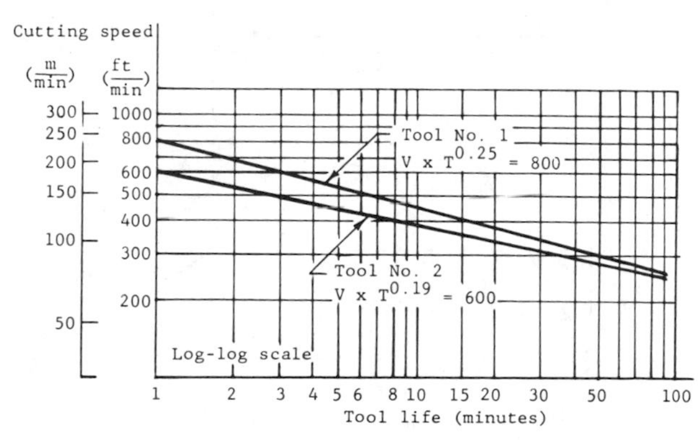

Fig. 5.17. Relationship between tool life and cutting speed for two tools.

the cost and production rate as a function of cutting speed can be calculated for tool 2 by applying Eqs. 5.27 and 5.13. The results for the production rate are shown in Fig. 5.18, where the comparative relationship between tools 1 and 2 is graphically illustrated.

The analytical expression for the production rate as a function of cutting speed for tool 2 can be determined by substituting numerical values into Eq. 5.13:

$$P_r = \frac{12 \times C \times f \times d \times T^{1-n}}{T + t_1}$$

$$= \frac{12 \times 600 \times 0.007 \times 0.1 \times T^{1-0.19}}{T+2}$$

$$= \frac{5.04 \times T^{0.81}}{T+2}$$

where for tool 2, $T = (600/V)^{1/n} = (600/V)^{5.26}$. Substituting yields

$$P_r = \frac{5.04[(600/V)^{5.26}]^{0.81}}{(600/V)^{5.26} + 2} \tag{5.41}$$

The cutting speed for maximum production for tool 2 can be evaluated from Eqs. 5.21 and Eq. 5.40:

$$T_{\mathrm{mp}} = \left(\frac{1}{n} - 1\right) \times t_1 = \left(\frac{1}{0.19} - 1\right) \times 2$$

$$= 8.526 \text{ min}$$

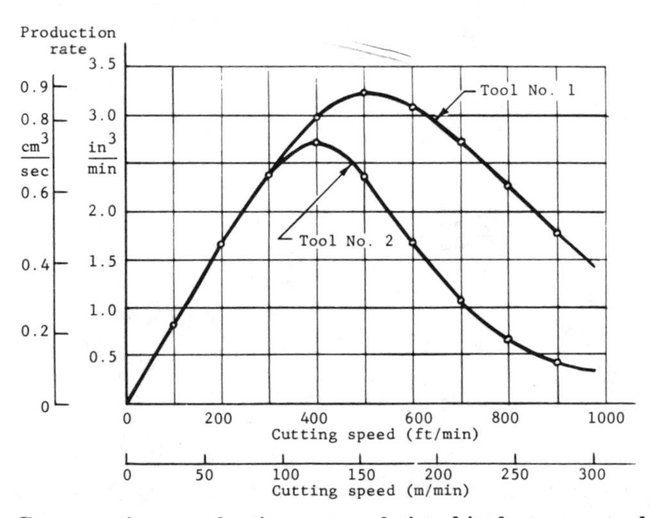

Fig. 5.18. Comparative production rate relationship between tools 1 and 2.

Substituting into equation 5.40 yields

$$V_{mp} = \frac{600}{T^{0.19}} = \frac{600}{(8.526)^{0.19}}$$

$$= 399.3 \text{ ft/min } (121.7 \text{ m/min})$$

With the cutting speed for maximum production available, the maximum production rate can be determined by substituting into Eq. 5.41. The result is

$$P_{rmp} = 2.717 \text{ in.}^3\text{/min } (0.7417 \text{ cm}^3\text{/sec})$$

As can be seen in Fig. 5.18, the production rate for both tools is essentially the same below a cutting speed of 300 ft/min (91 m/min). Above this point, the difference in production rate for the two tools becomes significant.

EXAMPLE 5.11. From the data used to generate Fig. 5.18, calculate the production rate for tool 2 for the same cutting speed that yields a maximum production for tool 1. As previously determined, the maximum production rate for tool 1 is 3.22 in.^{3}/min (0.879 cm^3/sec) at a cutting speed of 511 ft/min (155.8 m/min).

Taking into account the tool life equation for tool 2, namely,

$$V \times T^{0.19} = 600$$

and substituting numerical values into Eq. 5.13, we obtain

$$P_r = \frac{12 \times C \times f \times d \times T^{1-n}}{T + t_1}$$

where

$$T = \left(\frac{600}{V}\right)^{1/n} = \left(\frac{600}{511}\right)^{5.263}$$

$$= 2.328 \text{ min}$$

Thus

$$P_r = \frac{12 \times 600 \times 0.007 \times 0.100 \times (2.328)^{0.81}}{2.328 + 2}$$

$$= 2.309 \text{ in.}^3\text{/min } (0.6303 \text{ cm}^3\text{/sec})$$

The total cost per part as a function of cutting speed for tools 1 and 2 is shown in Fig. 5.19. In examining the cost relationship below 300 ft/min (91 m/min), the cost per part is essentially the same. Beyond this point, the difference in cost between the two tools becomes significant.

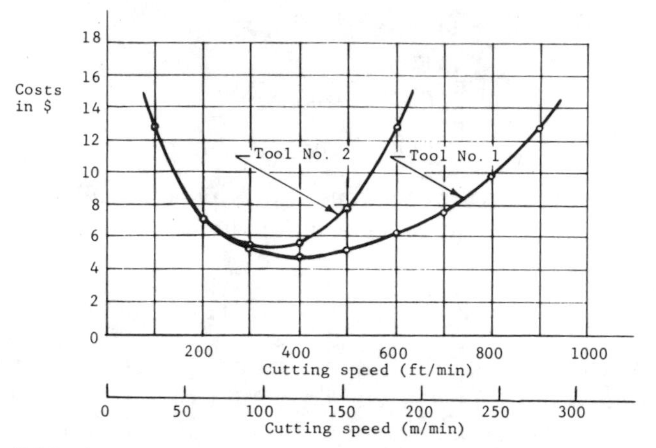

Fig. 5.19. Comparative cost relationship between tools 1 and 2.

EXAMPLE 5.12. Determine the minimum cost for tool 2 as shown in Fig. 5.19.

Solving for the tool life and cutting speed for minimum cost by substituting into Eqs. 5.29 and 5.40 yields

$$T_{mc} = \left(\frac{1}{n} - 1\right)\left(t_1 + \frac{C_e}{R_c}\right)$$

$$= \left(\frac{1}{0.19} - 1\right)\left(2 + \frac{1.75}{0.538}\right)$$

$$= 22.4 \text{ min}$$

and

$$V_{mc} = \frac{600}{(T_{mc})^n} = \frac{600}{(22.39)^{0.19}}$$

$$= 332.4 \text{ ft/min } (101.31 \text{ m/min})$$

Solving for the minimum cost by substituting into Eq. 5.28 yields

$$C_{Tmc} = \left(\frac{\pi \times D \times L}{12 \times f \times V_{mc}}\right)\left(R_c + \frac{R_c \times t_1}{T_{mc}} + \frac{C_e}{T_{mc}}\right) + (R_c \times t_n)$$

$$= \left(\frac{3.14 \times 3 \times 18}{12 \times 0.007 \times 322.4}\right)\left(0.583 + \frac{0.583 \times 2}{22.4} + \frac{1.75}{22.4}\right) + (0.583 \times 1.8)$$

$$= \$5.51$$

EXAMPLE 5.13. From the information provided in Fig. 5.19, calculate the difference in cost between tools 1 and 2 for the minimum-cost cutting-speed setting for tool 1. From previous calculations as shown in example 5.6, the minimum cost for tool 1 occurs when the tool life is 14.4 min with a corresponding cutting speed of 410.67 ft/min (125.17 m/min).

The minimum cost for tool 1 can be evaluated from Eq. 5.28 as

$$C_{\mathrm{Tmc}} = \left(\frac{\pi \times D \times L}{12 \times f \times V_{\mathrm{mc}}}\right)\left(R_c + \frac{R_c \times t_1}{T_{\mathrm{mc}}} + \frac{C_e}{T_{\mathrm{mc}}}\right) + (R_c \times t_n)$$

$$= \left(\frac{3.14 \times 3 \times 18}{12 \times 0.007 \times 410.67}\right)\left(0.583 + \frac{0.583 \times 2}{14.4} + \frac{1.75}{14.4}\right) + (0.583 \times 1.8)$$

$$= \$4.91$$

The tool life for tool 2 at a cutting speed setting of 410.67 ft/min (125.17 m/min) is

$$T = \left(\frac{600}{V}\right)^{1/n} = \left(\frac{600}{410.67}\right)^{5.263}$$

$$= 7.355 \text{ min}$$

Substituting these values into Eq. 5.28 for tool 2 yields

$$C_T = \left(\frac{\pi \times D \times L}{12 \times f \times V}\right)\left(R_c + \frac{R_c \times t_1}{T} + \frac{C_l}{T}\right) + (R_c \times t_n)$$

$$= \left(\frac{3.14 \times 3 \times 18}{12 \times 0.007 \times 410.67}\right)\left(0.583 + \frac{0.583 \times 2}{7.355} + \frac{1.75}{7.355}\right) + (0.583 \times 1.8)$$

$$= \$5.86$$

The difference in cost between tools 2 and 1 is

$$C_T = C_{t2} - C_{t1}$$

$$= 5.86 - 4.91$$

$$= \$0.95$$

It is possible to have two tools made of the same material where the tools have the same tool life equation but give different production rates and different costs when they are used. The case in question is the one comparing the throw-away insert versus the fixed insert (brazed) that can be reground. Usually, the throw-away insert has a lower cost per edge as well as a lower tool change time than the tool with the fixed insert that can be reground. In many cases, the reason

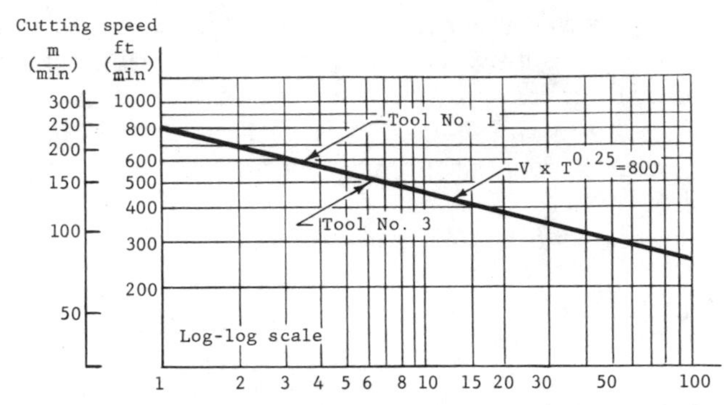

Fig. 5.20. Tool life as a function of cutting speed for tools 1 and 3.

the fixed insert is more expensive to use is not only because the regrinding is costly, but also because it takes more time to change the tool due to required positioning. A comparison of these two tools will be made to see how the cost and production rate is affected.

The case to be considered utilizes the same tool 1 that was used in the previous example. A comparison will be made with a fixed insert tool, labeled tool 3. For the sake of the example, everything will be the same between tools 1 and 3, with the exception of the tool change time and the cost per edge. For tool 3, the tool change time will be 5 min and the cost per edge will be \$5.00. With these values, comparative calculations can be made. The tool life graph is shown in Fig. 5.20, and production rates for the two tools are shown in Fig. 5.21. Figure 5.22 illustrates the difference in costs between the two tools.

An examination of the production rate and cost associated with the use of tool 3 as a function of cutting speed reveals that there is a relatively small difference below 300 ft/min (91.44 m/min) when compared to tool 1. This is a similar situation as that of tool 2. In both cases, the difference above 300 ft/min (91.44 m/min) is significant.

Of special interest is a comparison between tools 2 and 3. The production rate and cost are found to be relatively close. However, the tools possess significantly different features. These are concentrated primarily in the areas of tool life, the cost per cutting edge, and the tool change time. The main reason that the production rates and costs are relatively close is that the advantage that tool 3 has over tool 2 in tool life is balanced against a disadvantage in cost per edge as well as in tool change time. Table 5.5 lists the main features of tools 1, 2, and 3 with regard to turning the workpiece described earlier.

The importance of the tool cost per edge, as well as of the tool change time on maximum production and on minimum cost, is ap-

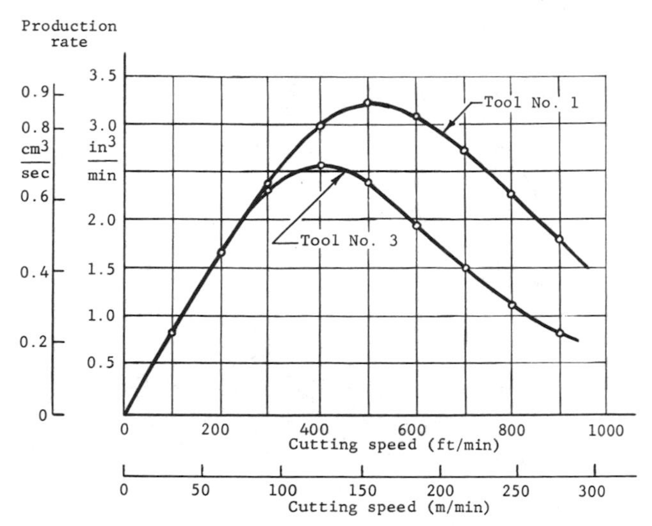

Fig. 5.21. Comparative production rate relationship between tools 1 and 3.

parent from Table 5.5. This is shown when comparing tool 1 with tool 3. These two tools have the same tool life equation but different values for tool change time and for cost per edge. On the other hand, the importance of tool life on maximum production and on minimum cost is also apparent in Table 5.5. By comparing tool 1 with tool 2, it can be seen that a shorter tool life has a tendency to reduce the production rate and to increase the cost per part of the operation.

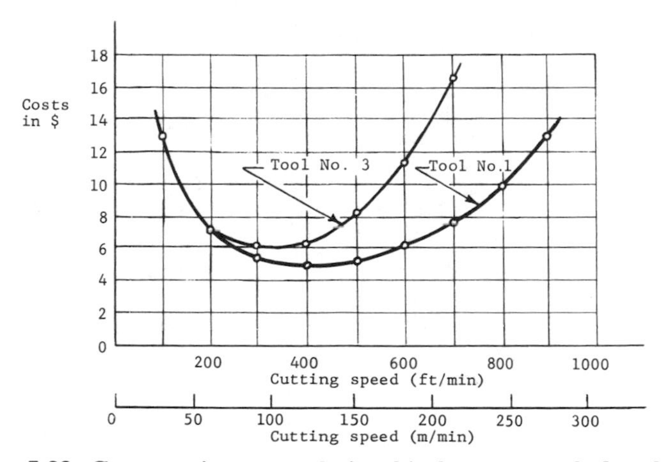

Fig. 5.22. Comparative cost relationship between tools 1 and 3.

Table 5.5. Main Features of Turning Example for Tools 1, 2, and 3

Feature	Tool 1	Tool 2	Tool 3
Tool life equation	$V \times T^{0.25} = 800$	$V \times T^{0.19} = 600$	$V \times T^{0.25} = 800$
Cost per edge	$1.75	$1.75	$5.00
Tool change time	2 min	2 min	5 min
Tool life for maximum production	6 min	8.5 min	15 min
Tool life for minimum cost	15 min	22.4 min	40.7 min
Cutting speed for maximum production	511 ft/min (156 m/min)	399 ft/min (122 m/min)	407 ft/min (124 m/min)
Cutting speed for minimum cost	406.5 ft/min (125 m/min)	332 ft/min (101 m/min)	316.7 ft/min (96.5 m/min)
Maximum production in parts per hour	8.71	7.65	7.3
Minimum cost	$4.91/part	$5.51/part	$6.00/part

EXAMPLE 5.14. For the turning example listed in Table 5.5 using tool 3, confirm the values indicated for:

(a) Tool life for maximum production.
(b) Tool life for minimum cost.
(c) Cutting speed for maximum production.
(d) Cutting speed for minimum cost.
(e) Maximum production in parts per hour.
(f) Minimum cost per part.

The tool life equation is given as

$$V \times T^{0.25} = 800$$

Other values include a tool cost per edge of $5.00 and a tool change time of 5 min. The machine settings are 0.007 in./rev (0.1778 mm/rev) for feed and 0.100 in. (2.54 mm) for depth of cut. The turning length is 18 in. (457.2 mm) and the outside diameter of the part is 3 in. (76.2 mm). The charge rate is $0.583/min and the nonproductive time is 1.8 min/part.

The tool life for maximum production can be evaluated from Eq. 5.21 as

$$T_{mp} = t_1\left(\frac{1}{n} - 1\right) = 5\left(\frac{1}{0.25} - 1\right)$$
$$= 15 \text{ min}$$

The tool life for minimum cost can be evaluated from Eq. 5.29 as

$$T_{mc} = \left(\frac{1}{n} - 1\right)\left(t_1 + \frac{C_e}{R_c}\right) = \left(\frac{1}{0.25} - 1\right)\left(5 + \frac{5.00}{0.583}\right)$$
$$= 40.7 \text{ min}$$

By substituting into the tool life equation for tool 3, the cutting speed for maximum production can be written as

$$V_{mp} = \frac{800}{(T_{mp})^{0.25}} = \frac{800}{(15)^{0.25}}$$
$$= 406.5 \text{ ft/min } (123.9 \text{ m/min})$$

The cutting speed for minimum cost, in turn, can be written as

$$V_{mc} = \frac{800}{(T_{mc})^{0.25}} = \frac{800}{(40.7)^{0.25}}$$
$$= 316.7 \text{ ft/min } (96.5 \text{ m/min})$$

The maximum production rate can be evaluated from Eq. 5.13 by substituting the tool life for maximum production. As a result,

$$P_{\text{rmp}} = \frac{12 \times C \times f \times d \times T_{\text{mp}}^{1-n}}{T_{\text{mp}} + t_1}$$

$$= \frac{12 \times 800 \times 0.007 \times 0.100 \times 15^{(0.75)}}{15 + 5}$$

$$= 2.56 \text{ in.}^3/\text{min } (0.699 \text{ cm}^3/\text{min})$$

In terms of parts per hr, the production rate can be evaluated from Eq. 5.33, where the volume per cut is 16.4 in.^{3} (268.6 cm^3) as determined in example 5.9. As a result,

$$P_h = \frac{60 \times P_r}{V_L + (P_r \times t_n)} = \frac{60 \times 2.56}{16.4 + (2.56 \times 1.8)}$$

$$= 7.3 \text{ parts/hr}$$

Finally, the minimum cost per part can be evaluated from Eq. 5.28 by substituting values for the tool life at minimum cost and for the cutting speed at minimum cost. As a result,

$$C_{\text{Tmc}} = \left(\frac{\pi \times D \times L}{12 \times f \times V_{\text{mc}}}\right)\left(R_c + \frac{R_c \times t_1}{T_{\text{mc}}} + \frac{C_e}{T_{\text{mc}}}\right) \times (R_c \times t_n)$$

$$= \left(\frac{3.14 \times 3 \times 18}{12 \times 0.007 \times 316.7}\right)\left(0.583 + \frac{0.583 \times 5}{40.7} + \frac{5.00}{40.7}\right)$$

$$\times (0.583 \times 1.8)$$

$$= \$6.00$$

An interesting point in cost analysis is an evaluation of the expenditure for energy. Usually this value is low when compared with labor costs and tooling costs. The energy cost at the cut, neglecting energy losses due to machine efficiency as well as energy losses due to tare power, can be expressed as

$$E_c = \frac{E_r \times U_p \times V_L}{K} \quad \textbf{(5.42)}$$

where E_r is the energy rate (\$/kW hr), U_p is the unit horsepower (hp/in.^{3}/min), V_L is the volume cut (in.^{3}), and K is the unit balance factor (80.4 hp-min/kW-hr). Example 5.15 illustrates the evaluation of the energy cost at the cut for the turning example listed in Table 5.5.

ot the case over a long period of time, where the interest charge be substantial. Figure 5.23 illustrates the case of a tool investment h no interest charge attached to it.

As can be seen, the tool cost remains constant whereas the income roduced by the tool investment is a straight-line linear function. The otal income from the use of the tool can be written as

$$T_i = C \times t \tag{5.48}$$

where T_i is the total income from use of tool, t is the time, and C is the income produced by the tool per unit time. When the income produced by the tool investment is equal to the amount invested to purchase the tool, then the break-even point is reached. This is the point where the tool cost is recovered and when the tool begins to produce a gain on its investment. The break-even time can be written as

$$T_i = C \times t_{\text{be}} = T_c$$

or

$$t_{\text{be}} = \frac{T_c}{C} \tag{5.49}$$

where T_c is the tool cost, t_{be} is the break-even time, and C is the income produced by the tool per unit time.

EXAMPLE 5.19. A tool expenditure of $1000 is to be made for a tool that is expected to make a contribution toward profitability of $120.00 per day while it is being used. If the tool has a useful life of 25 days, determine the break-even time for use of the tool as well as the gain that the tool will produce through its use.

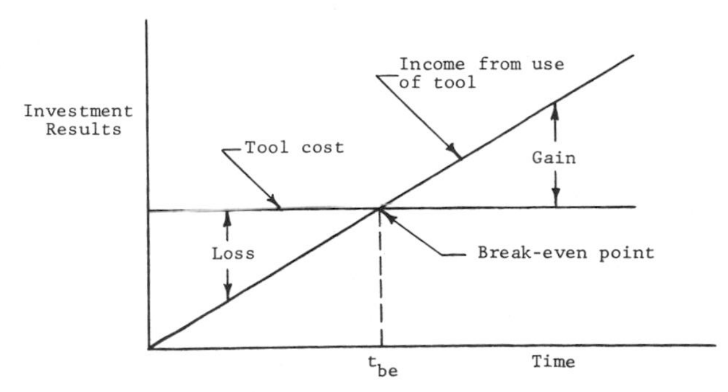

Fig. 5.23. Tool cost and income as a function of time without the influence of interest.

EXAMPLE 5.15. Determine the cost of the energy expended at the cut for a turning case where the volume of material removed is 16.4 in.^{3}, the unit horsepower for the work material is 0.85 hp/in.^{3}/min, and the energy rate is $0.06/kW-hr.

Substituting into Eq. 5.42 yields

$$E_c = \frac{E_r \times U_p \times V_L}{K} = \frac{0.06 \times 0.85 \times 16.4}{80.4}$$

$$= \$0.0104$$

5.7 FINANCIAL CONSIDERATIONS

The time rate associated with borrowing money is an important factor in the consideration of tooling expenditures. If a tool is to be purchased, then the money spent on the tool is no longer available to produce income. Instead, the tool is expected to make a profit, through use, which includes recovery of the loss of income on the money spent. On the other hand, if money is not available to pay for a tooling expenditure, then financing is usually arranged. In this case, the tool must bear the burden of the financial charge. In either case, a tooling expenditure is expected to earn a return on its investment that is above the finance charge or the time rate. Should this not be the case, then the expenditure would yield a net loss and the investment would not be justified.

The use of money is associated with the term *interest*. In banking, interest is defined as the rate of return on capital that is paid to a person who has money deposited in a bank. In the case of a loan, interest, the time rate associated with borrowing money, is then charged to the borrower for the use of the money.

In the case where the interest is left on deposit, the interest has interest paid on it. The payment of interest on interest along with the interest on the original deposit is known as *compound interest*. An analytical expression that describes the results of compound interest is given by

$$F = P(1 + I)^n \tag{5.43}$$

where P is the present sum, F is the future sum, I is the interest rate, and n represents the number of interest periods. The number of interest periods has an effect on the future sum of money as a result of interest being paid on interest. This is illustrated by the following example.

EXAMPLE 5.16. An amount of \$10,000 is left on deposit for a 5-yr period at an annual interest rate of 7.5%. Compare the difference between annual compounding and monthly compounding on the future sum.

For annual compounding, $I_p = 0.075$ and $n = 5$. Substituting into Eq. 5.43 yields

$$F_1 = P(1 + I)^n = 10{,}000(1 + 0.075)^5$$
$$= \$14{,}356.29$$

For monthly compounding, $I_p = 0.075/12 = 0.00625$ and $n = 5 \times 12 = 60$. Substituting into Eq. 5.43 yields

$$F_2 = 10{,}000(1 + 0.00625)^{60}$$
$$= \$14{,}532.94$$

The difference can be written as

$$\Delta F = F_2 - F_1 = 14{,}532.94 - 14{,}356.29$$
$$= \$176.65$$

If a fixed sum is invested periodically for a specific purpose, it is referred to as an *annuity*. The term *sinking fund* is used for an annuity that has as a purpose the generation of sufficient funds at the end of a given period to repay the principal of a debt or to provide a definite sum for some other purpose. The future sum of an annuity or of a sinking fund can be expressed as

$$F = R\left(\frac{(1 + I)^n - 1}{I}\right) \tag{5.44}$$

where F is the future sum, R is the amount set aside, I is the interest rate, and n represents the number of interest periods.

EXAMPLE 5.17. A young tool engineer has started a tax-sheltered retirement account with a \$150 monthly investment at an annual return of 9%. What will the annuity be worth in 40 years?

Substituting into Eq. 5.44 yields

$$F = R\left[\frac{(1 + I)^n - 1}{I}\right] = (150 \times 12)\left[\frac{(1.09)^{40} - 1}{0.09}\right]$$
$$= \$608{,}188.40$$

In the case of installment buying, where a loan is taken out to be repaid (principal plus interest) in a certain period, the periodic pay-

ment can be written as

$$R = \frac{A \times I}{1 - [1/(1 + I)]^n}$$

where A is the amount of loan, I is the interest ra[...] payment, and n represents the number of interest [...] amount repaid in an installment loan can be written a[...]

$$A_p = R \times n$$

This amount is obviously greater than the amount of the [...] includes payments on the principal as well as payments on t[...] of the unpaid balance. The difference between the total amo[...] on the loan and the amount of the loan is equal to the total i[...] payments. This can be expressed as

$$P_i = A_p - A \quad (5.$$

Example 5.18 illustrates an application of Eqs. 5.45–5.47.

EXAMPLE 5.18. A \$15,000 automobile loan is to be repaid in monthly payments over a 4-yr period. The annual interest rate is set at a 7%. Determine the monthly payment as well as the total amount that is paid on the loan. In addition, calculate the amount of the total interest payments.

Substituting into Eq. 5.45 yields

$$R = \frac{A \times I}{1 - [1/(1 + I)]^n}$$
$$= \frac{15{,}000 \times (0.07/12)}{1 - \{1/[1 + (0.07/12)]\}^{48}}$$
$$= \$359.19/\text{month}$$

The total amount paid on the loan is determined from Eq. 5.46 as

$$A_p = R \times n = 359.19 \times 48$$
$$= \$17{,}241.12$$

The total amount paid for interest can be evaluated from Eq. 5.47 as

$$P_i = A_p - A = 17{,}241.12 - 15{,}000.00$$
$$= \$2241.12$$

Since the interest charge is a function of time, economic evaluation over a very short period of time is not influenced by the interest. This

Solving for the break-even time by substituting into Eq. 5.49 yields

$$t_{be} = \frac{T_c}{C} = \frac{1000}{120}$$
$$= 8.33 \text{ days}$$

The total gain that can be expected from the use of the tool is the difference between the income produced by the tool and the cost of the tool, that is,

$$\begin{aligned} \text{Gain} &= T_i - T_c = (C \times t) - T_c \\ &= (120 \times 25) - 1000 \\ &= \$2000 \end{aligned}$$

When interest is taken into consideration over a long period of time, the relationship between tool cost and total income as a function of time is of the form shown in Fig. 5.24. As can be seen, the tool cost takes into account the lost income that would have been produced by the present sum (P) that has been invested in the tool. The income produced through the use of the tool is also influenced by interest accumulation, assuming that the income is invested. The tool cost can then be expressed in the form of Eq. 5.43 as

$$T_c = F_c = P(1 + I)^n \tag{5.50}$$

and the total income from the use of the tool can be expressed in the form of Eq. 5.44 as

$$T_i = F_i = R\left(\frac{(1 + I)^n - 1}{I}\right) \tag{5.51}$$

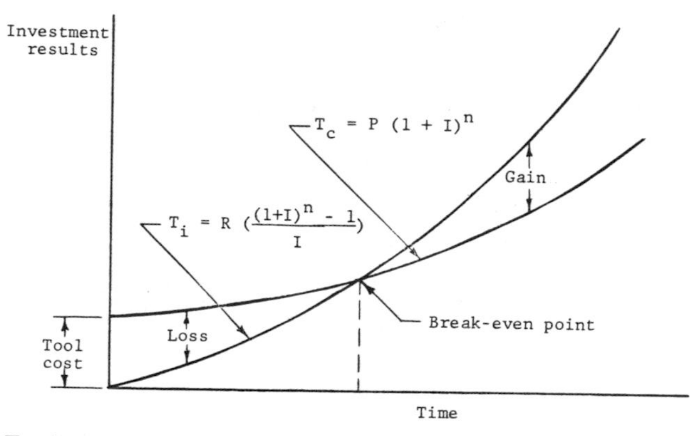

Fig. 5.24. Tool cost and tool income as a function of time, as influenced by interest rates.

where P is the present sum invested in tool, R is the periodic income produced by tool, n is the number of periods, and I is the interest rate. The break-even point in terms of the number of periods (time) can be evaluated by equating Eq. 5.50 with Eq. 5.51 and solving for n. As a result,

$$T_c = T_i$$

$$P(1 + I)^n = R\left[\frac{(1 + I)^n - 1}{I}\right]$$

$$(P \times I)(1 + I)^n = R(1 + I)^n - R$$

$$(1 + I)^n = \frac{-R}{(P \times I) - R} = \frac{R}{R - (P \times I)}$$

$$n \log(1 + I) = \log \frac{R}{R - (P \times I)}$$

$$n = \frac{\log\{R/[R - (P \times I)]\}}{\log(1 + I)} \tag{5.52}$$

EXAMPLE 5.20. A machine tool expenditure of \$250,000 is to be paid for by money that is earning interest at a rate of 9% annually. The machine tool is expected to provide an increase of income of \$75,000/yr, which is to be invested at the same interest rate of 9% that is earned by the money spent on the machine tool. Taking into account the interest lost on the amount spent on the machine as well as the interest earned on the increased income that is to be invested, determine:

(a) The break-even time for the investment to pay for itself.
(b) The total gain if the machine has a useful life of 10 yr and the interest rate is assumed to remain constant.

Solving for the number of yearly periods to reach the break-even point by substituting into Eq. 5.52 yields

$$n = \frac{\log\{R/[R - (P \times I)]\}}{\log(1 + I)}$$

$$= \frac{\log\{75{,}000/[75{,}000 - (250{,}000 \times 0.09)]\}}{\log 1.09}$$

$$= \frac{0.1549}{0.0374}$$

$$= 4.139 \text{ yr}$$

Table 5.6. Listing of Investment Results

Period (Years)	Tool Cost		Tool Investment Return	
	Without Interest	With Lost Interest	Without Interest	With Interest
1	250,000	272,500	75,000	75,000
2	250,000	297,025	150,000	156,750
3	250,000	323,757	225,000	245,858
4	250,000	352,895	300,000	342,985
5	250,000	389,656	375,000	448,853
6	250,000	419,275	450,000	564,250
7	250,000	457,010	525,000	690,032
8	250,000	498,140	600,000	827,136
9	250,000	542,973	675,000	976,578
10	250,000	591,841	750,000	1,139,470

The total gain for a 10-yr use period can be written in terms of Eqs. 5.50 and 5.51 as

$$\begin{aligned}
\text{Gain} &= F_i - F_c \\
&= R\left[\frac{(1+I)^n - 1}{I}\right] - P(1+I)^n \\
&= 75{,}000\left[\frac{(1.09)^{10} - 1}{0.09}\right] - 250{,}000(1.09)^{10} \\
&= 1{,}139{,}469.73 - 591{,}840.91 \\
&= \$547{,}628.82
\end{aligned}$$

The numerical values for example 5.20 are listed in Table 5.6 and are plotted in Fig. 5.25.

EXAMPLE 5.21. If the $250,000 payment for the machine tool listed in example 5.20 is taken as a loan at a 10% annual rate for a 10-yr period, to be repaid in yearly installments, determine the amount of the yearly payment.

Substituting numerical values into Eq. 5.45 yields

$$\begin{aligned}
R &= \frac{A \times I}{1 - [1/(1+I)]^n} \\
&= \frac{250{,}000(0.10)}{1 - [1/(1.1)]^{10}} \\
&= \$40{,}686.35
\end{aligned}$$

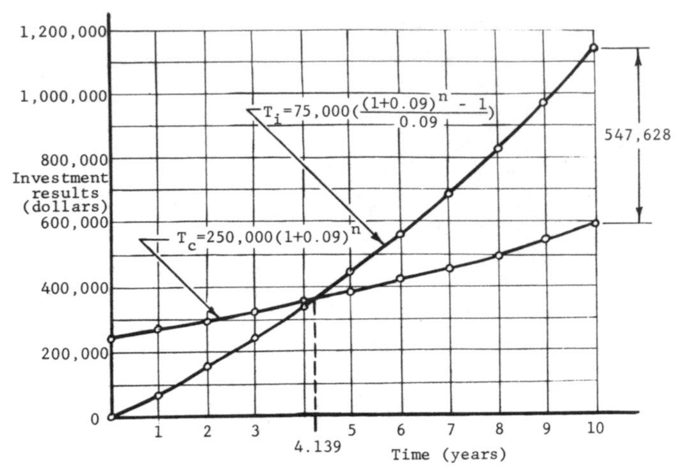

Fig. 5.25. Graphical representation of investment result.

5.8 SUMMARY OF TOOL WEAR AND AFFILIATED PRODUCTION COSTS

When a tool loses its capability to produce parts within given specifications, it is considered to have failed. Usually, the failure is due to wear on the tool. In the process of cutting metal, a tool is subjected to the harsh environment of high temperature and high contact forces. This condition leads to a situation where the tool is continuously sacrificing itself in the performance of the cutting process. The progression of the wear is a complex event that is influenced by temperature. Experimental evidence indicates that tool life, which is directly related to the wear of the tool, decreases with an increase in cutting speed.

Tests can be conducted to confirm the relationship between tool life and cutting speed. From the test results, an analytical expression can be derived that describes the relationship between tool life and cutting speed for a given tool machining a particular work material. Once a relationship is established between tool life and cutting speed, production rates can be analytically written in terms of tool life. By differentiating the production-rate tool life expression and setting the slope (derivative) at zero, the tool life for maximum production can be determined. This leads to the optimization of cutting speed for a maximum production case.

In a similar fashion, with costs expressed in terms of tool life, an expression can be derived to optimize machining conditions in terms of a minimum-cost operation. By comparing conditions for maximum production and for minimum cost, an idealized operating range can be defined. This is a range where high production rates exist at corresponding low-cost conditions.

A comparison of the performance of different tools for cost and for production can be revealing. If two tools possess a different tool-life–cutting-speed relationship, costs and production results in the higher speed range will also be different. On the other hand, if there is a difference in tool change time or in the cost of the tool, even if the tool-life–cutting-speed relationship is the same for the two tools, the performance in terms of cost and production can be shown to be different.

Finally, a financial consideration of the time rate of an investment is shown to have an influence on a tooling expenditure. The tool must bear the burden of the financial charge in terms of the contribution it is expected to make toward profitability. If a tooling investment does not return at least the time rate (interest) that the expenditure can earn, then justifying the expenditure comes under question since it will produce a net financial loss.

PROBLEMS

PROBLEMS 5.1. Two tests are conducted to determine the relationship between cutting speed and tool life. In the first test, a cutting speed of 300 ft/min results in a tool life of 25 min. In the second test, a cutting speed of 200 ft/min results in a tool life of 65 min. From these data, determine the constants in Taylor's tool life equation.

Answer: $V \times T^{0.424} = 1174$

PROBLEM 5.2. Two tests are conducted using SI units to determine the relationship between cutting speed and tool life. The test results are the same as in problem 5.1. In the first test, a cutting speed of 91.4 m/min results in a tool life of 25 min. In the second test, a cutting speed of 60.96 m/min results in a tool life of 65 min. From these data, determine the constants in Taylor's tool life equation.

Answer: $V \times T^{0.424} = 357.65$

PROBLEM 5.3. Using the tool life equation from problem 5.1, calculate the expected tool life for a cutting speed setting of 250 ft/min.

Answer: $T = 38.37$ min

PROBLEM 5.4. Using the tool life equation from problem 5.2, calculate the expected tool life for a cutting speed setting of 76.2 m/min.

Answer: $T = 38.35$ min

PROBLEM 5.5. The tool life for a machining operation can be expressed in U.S. customary units as

$$T = \frac{50 \times 10^6}{V^{4.2} \times f^{1.5} \times d^{0.5}}$$

Using this expression, determine the tool life for the following machine settings:

$$\text{Cutting speed} = 200 \text{ ft/min}$$
$$\text{Feed} = 0.010 \text{ in./rev}$$
$$\text{Depth of cut} = 0.060 \text{ in.}$$

Answer: $T = 44.2$ min

PROBLEM 5.6. The tool life for a machining operation can be expressed in SI units as

$$T = \frac{219.44 \times 10^6}{V^{4.2} \times f^{1.5} \times d^{0.5}}$$

Using this expression, determine the tool life for the following machine settings:

$$\text{Cutting speed} = 60.96 \text{ m/min}$$
$$\text{Feed} = 0.254 \text{ mm/rev}$$
$$\text{Depth of cut} = 1.524 \text{ mm}$$

Answer: $T = 44.2$ min

PROBLEM 5.7. Using the expressions and data from either problem 5.5 or problem 5.6, calculate the tool life for:

(a) A 25% increase in cutting speed.
(b) A 25% increase in feed.
(c) A 25% increase in depth of cut.

Answer: (a) With $V_2 = 1.25V_1$, $T = 17.3$ min
(b) With $f_2 = 1.25f_1$, $T = 31.6$ min
(c) With $d_2 = 1.25d_1$, $T = 39.5$ min

PROBLEM 5.8. Determine the volume of metal cut by a single edge of a tool that has a tool life of 60 min. The cutting speed is 50 m/min, the feed is 0.150 mm/rev, and the depth of cut is 1.600 mm.

Answer: $Q_t = 720$ cm^3

PROBLEM 5.9. For the case shown in Fig. 5.10, where the tool life equation is of the form

$$V \times T^{0.25} = 800$$

the feed is 0.007 in./rev, the depth of cut is 0.100 in., and the tool change time is 2 min, determine the production rate for a tool life of 30 min.

Answer: $P_r = 2.69\ \text{in.}^3/\text{min}$

PROBLEM 5.10. In SI units, for the case shown in Fig. 5.10, the tool life equation can be written as

$$V \times T^{0.25} = 243.8$$

where the cutting speed (V) is given in units of meters per minute. Using the same data as in problem 5.9, namely,

$$f = 0.178\ \text{mm/rev},\ d = 2.54\ \text{mm, and}\ t_1 = 2\ \text{min}$$

determine the production rate for a tool life of 30 min.

Answer: $P_r = 0.7359\ \text{cm}^3/\text{sec}$

Note: In SI units, Eq. 5.13 is written as

$$P_r = \frac{C \times f \times d \times T^{(1-n)}}{60(T + t_1)}$$

PROBLEM 5.11. Given the tool life equation in U.S. customary units as

$$V \times T^{0.25} = 800$$

determine the tool life and cutting speed for maximum production for a case where the tool change time is 2 min.

Answers: $T_{mp} = 6\ \text{min}$
$V_{mp} = 511\ \text{ft/min}$

PROBLEM 5.12. Given the tool life equation in *SI* units as

$$V \times T^{0.25} = 243.8$$

determine the tool life and cutting speed for maximum production for the case where the tool change time is 2 min.

Answers: $T_{mp} = 6\ \text{min}$
$V_{mp} = 155.8\ \text{m/min}$

PROBLEM 5.13. Confirm the cost data in Table 5.4 for the turning example with a cutting speed setting of 200 ft/min. The tool life equation is

$$V \times T^{0.25} = 800$$

The length of cut is 18 in. on a 3-in.-diameter shaft with a feed of 0.007 in./rev and a depth of cut of 0.100 in. The tool change time is 2 min and the nonproductive time per piece (loading, etc.) is 1.8 min. The charge rate is \$0.583/min and the tooling cost is \$1.75 per edge. From these data determine:

(a) Cutting cost.
(b) Tool change cost.
(c) Tool cost.
(d) Nonproduction cost.
(e) Total cost.

Answers: $C_c = \$5.88$
$C_h = \$0.046$
$C_t = \$0.069$
$C_n = \$1.049$
$C_T = \$7.05$

PROBLEM 5.14. Determine the cutting cost (C_c) for a turning case where the diameter is 76.2 mm, the length of cut is 457.2 mm, the feed setting is 0.1778 mm/rev, and the cutting speed is 60.96 m/min. The charge rate is \$0.583/min.

Answer: $C_c = \$5.88$

Note: In SI units, Eq. 5.23 is written as

$$C_c = \frac{R_c \times L \times \pi \times D}{f \times V \times 10^3}$$

PROBLEM 5.15. A metal cutting operation is described by

$$V \times T^{0.18} = 400$$

where V is given in units of feet per minute. If the tool has a change time of 1 min and a cost of \$1.50 per edge, calculate the tool life and cutting speed for a minimum-cost operation. The charge rate is \$0.60/min.

Answers: $T_{\mathrm{mc}} = 15.94$ min
$V_{\mathrm{mc}} = 243$ ft/min

PROBLEM 5.16. Calculate the production rate for minimum cost for the case illustrated in problem 5.15 for a feed setting of 0.127 mm/rev and a depth of cut of 1.524 mm.

Answer: $P_{\mathrm{rmc}} = 0.225\ \mathrm{cm}^3/\mathrm{sec}$

PROBLEM 5.17. Determine the production rate in parts per hour for a turning case where the tool life is 30 min and the tool life equation is of the form

$$V \times T^{0.21} = 350$$

The part has a diameter of 2 in. and is being turned for a length of 10 in. The feed is set at 0.008 in./rev, the depth of cut is set at 0.075 in, and the tool change time is 1.5 min. The nonproduction time (loading, etc.) is equal to 1 min.

In addition, calculate the volume per cut as well as the production rate in units of cubic inches per minute.

Answers: $V_L = 4.533$ in.^{3}
$P_r = 1.174$ in.^{3}/min
$P_h = 12.34$ parts/hr

PROBLEM 5.18. Calculate the volume of material removed per cut for the following turning example. The outside diameter of the part is 50 mm, the depth of cut is 2.0 mm, and the length of the cut is 250 mm.

Answer: $V_L = 75.36$ cm^3

PROBLEM 5.19. Calculate the energy cost for a metal cutting operation where 20 in.^{3} of material is removed from a workpiece that has a unit horsepower of 0.95 hp/in.^{3}/min. The energy rate is $0.07/kW-hr.

(a) Neglect losses due to machine efficiency and tare power.
(b) Consider total machine losses to be 30%.

Answers: (a) $E_c = \$0.0165$
(b) $E_c = \$0.0236$

PROBLEM 5.20. Determine the energy cost for a metal cutting operation where 327.74 cm^3 of material is removed from a work piece that has a unit power of 2.5935 kW/cm^3/sec. The energy rate is $0.07/kW-hr.

(a) Neglect losses due to machine efficiency and tare power.
(b) Consider total machine losses to be 30%.

Answers: (a) $E_c = \$0.0165$
(b) $E_c = \$0.0236$

Note: In SI units, Eq. 5.42 is written as

$$E_c = E_r \times U_p \times V_L$$

PROBLEM 5.21. Determine the future sum of an initial investment of $10,000 that is compounded monthly for a 10-yr period at an annual rate of 7.0%.

Answer: $F = \$20{,}088.62$

PROBLEM 5.22. A young tool engineer has aspirations of having his investment double in value over a 5-yr period. In order for this to transpire, what is the required annual compounded rate of growth?

Answer: $I = 14.87\%$

PROBLEM 5.23. If through the use of a machine tool $5000/month is earned and the earnings are invested at a 10% annual rate that is compounded monthly, determine the value of the investment at the end of a 5-yr period.

Answer: $F = \$387{,}144.43$

PROBLEM 5.24. A machine tool is purchased with installment loan financing. If the amount of the loan is $100,000 at an annual rate of 9.5%, to be repaid in monthly installments over a 5-yr period, determine the amount of the monthly payment.

Answer: $R = \$2{,}100.19$

PROBLEM 5.25. A tool expenditure of $1500 is to be made for a tool that is expected to increase profitability by $100/day while it is being used. It the tool has a useful life of 30 days, determine the break-even time for use of the tool as well as the gain that can be expected through use of the tool.

Answers: $t_{be} = 15$ days
Gain = $1500

PROBLEM 5.26. A machine tool expenditure of $300,000 is to be paid for with capital that is earning interest at a rate of 8% annually. The machine tool is expected to provide a direct increase in income through its use of $150,000/yr. This income is to be invested at the same interest rate of 8% that is currently being earned by the capital that is to be expended for the machine tool. Taking into account the interest lost on the amount spent on the machine tool as well as the interest earned on the increase income that is to be invested, determine:

(a) The break-even time for the investment to pay for itself.
(b) The total gain if the machine has a useful life of 8 yr and the interest rate is assumed to remain constant.

Answers: $n = 2.267$ yrs
Gain = $1,040,215.88

PROBLEM 5.27. A young entrepreneurial tool engineer expects to earn 25% on an investment in a new venture. He is able to take out a loan for $1,000,000. The annual interest rate on the loan is 9%. The principal ($1,000,000) is to be repaid at the end of 10 yr. He is to pay the interest charge (9%) annually and to repay the principal ($1,000,000) in a lump sum at the end of 10 yr (sinking fund).

With this financial arrangement, determine:

(a) The total interest paid on the loan.
(b) The annual amount that must be set aside for the sinking fund.
(c) The total amount paid in on the loan (sinking fund payments plus interest payments).

Answers: $T_I = \$900{,}000$
$R = \$30{,}072.56$
$T_a = \$1{,}100{,}725.60$

PROBLEM 5.28. Calculate the cutting speed for maximum production for the case of a high-speed tool machining a cast iron workpiece. The tool wear equation for the operation is given in the form

$$V \times T^{0.12} = 54.864$$

where the units for the cutting speed (V) are given in units of meters per minute. The tool change time is 1.5 min.

Answer: $V_{\mathbf{mp}} = 41.15$ m/min

PROBLEM 5.29. Determine the tool life and the cutting speed for minimum cost for the case where the tool life equation is of the form

$$V \times T^{0.19} = 300$$

The tool change time is 1.5 min, the cost per cutting edge is $1.60, and the charge rate is $0.50/min.

Answers: $\mathrm{T}_{\mathbf{mc}} = 20.04$ min
$V_{\mathbf{mc}} = 169.7$ ft/min

PROBLEM 5.30. A metal cutting operation is represented by the following tool life equation:

$$V \times T^{0.2} = 185$$

where the cutting speed (V) is given in units of meters per minute. If the cutting speed is increased 10% from 100 m/min to 110 m/min, determine the percentage change in the tool life.

Answers: $\Delta\mathrm{T}\% = 37.9\%$

BIBLIOGRAPHY

Bhattacharyya, Amitabka, and Ham, Inyong, *Design of Cutting Tools*, Society of Manufacturing Engineers, Dearborn, Michigan, 1969.

Blanchard, Benjamin S., *Engineering Organization and Management*, Prentice-Hall, Englewood Cliffs, New Jersey, 1976.

Boston, Orlan William, *Metal Processing*, John Wiley & Sons, New York, 1951.

Cook, Nathan H., *Manufacturing Analysis*, Addison-Wesley, Reading, Massachusetts, 1966.

Cook, Nathan H., "Tool Wear and Tool Life," Paper No. 72-WA/Prod-19, American Society of Mechanical Engineers, New York, 1973.

DeVries, Marvin F., *Machining Economics—A Review of the Traditional Approaches and Introduction to New Concepts*, Society of Manufacturing Engineers, Dearborn, Michigan, 1969.

DeVries, M. F., and Nassirpour, F., *Metal Cutting Theory—Simplified*, Society of Manufacturing Engineers, Dearborn, Michigan, 1973.

Ham, Inyong, "Economics of Machining: Analyzing Optimum Machining Condition by Computers," Technical Paper, Society of Manufacturing Engineers, Dearborn, Michigan, 1964.

Hoffman, Edward G., *Fundamentals of Tool Design*, Society of Manufacturing Engineers, Dearborn, Michigan, 1984.

Komanduri, Ranga, "High-Speed Machining," *Mech. Eng.* December 1985.

Machining Data Handbook, Metcut Research Associates, Cincinnati, Ohio, 1972.

Mayer, John E., Jr., "Cutting Tool Monitoring," *J. Comput. Manuf.* December 1985.

Metals Handbook, American Society of Metals, Metals Park, Ohio, 1967.

Subramanian, K., and Cook, N. H., *Sensing of Drill Wear and Prediction of Drill Life*, American Society of Mechanical Engineers, New York, 1976.

Taylor, F. W., *On the Art of Cutting Metals*, American Society of Mechanical Engineers, New York, 1906.

Taylor, George A., *Managerial and Engineering Economy*, Van Nostrand, New York, 1980.

Tee, Liong-Hian, Ho, Show-Chung, and DeVries, Marvin F., *Economic Machining Charts*, Society of Manufacturing Engineers, Dearborn, Michigan, 1969.

Index

Index

A

B

C

D

E

F

G

H

I

J

L

M

N

O

P

R

S

T

U

V

W

Y

Z